INFORMATION SYSTEMS FOR URBAN PLANNING

A hypermedia co-operative approach

Robert Laurini

London and New York

First published 2001
by Taylor & Francis
11 New Fetter Lane, London EC4P 4EE

Simultaneously published in the USA and Canada
by Taylor & Francis Inc,
29 West 35th Street, New York, NY 10001-2299

Taylor & Francis is an imprint of the Taylor & Francis Group

© 2001 Robert Laurini

Typeset in Sabon by
MHL Typesetting Limited, Coventry, Warwickshire
Printed and bound in Great Britain by
Biddles Ltd, Guildford and King's Lynn

Every effort has been made to ensure that the advice and information in
this book is true and accurate at the time of going to press. However,
neither the publisher nor the author can accept any legal responsibility or
liability for any errors or omissions that may be made. In the case of
drug administration, any medical procedure or the use of technical
equipment mentioned within this book, you are strongly advised to
consult the manufacturer's guidelines.

British Library Cataloguing in Publication Data
A catalogue record for this book is available
from the British Library

Library of Congress Cataloging in Publication Data

Laurini, Robert.
  Information systems for urban planning: a hypermedia co-operative
  approach / Robert Laurini
    p. cm.
  Includes bibliographical references and index.
  1. City planning – Data processing.   I. Title.

HT166.L383 2001
307.1'2'0286–dc21   00-057713
                         ISBN 0-7484-0963-7 (hbk)
                              0-7484-0964-5 (pbk)

CONTENTS

FOREWORD

Forward digital paradigms: planning in an information age

It is the year 2020. An urban planner in the strategy division of the BosWash metropolis, the authority established to preserve and plan for the growth and regeneration of America's largest urban region, is watching heavy traffic around one its major international airports. The screen constantly updates the picture and the relevant portion of the traffic database in real time while the sliders and dials which form the interface within which the picture is embedded, provide critical information about the long-term average growth in traffic in this area and its relationships to the wider picture at different metropolitan scales. A similar picture based on a related data bases is also being used for traffic control in another section of the authority where traffic managers can override the automated software which controls the various highway infrastructures which regulate the flow. The urban planner in the strategy division is charged with watching changes in land use and traffic in this specific area through accessing databases which are updated from countless sources: in real time from sensors which monitor all kinds of change from land use to traffic, from periodically refreshed databases which record development applications and permits, and from agencies who update local economic and demographic data from national and international online sources and censuses.

Elsewhere in the agency, this kind of data is being represented in a form which can be accessed by local policy-makers and elected representatives. Relevant information about what is happening to a range of issues concerning the citizenry is also being presented so that the wider public can learn to interpret the impact of ongoing and perpetual change to their environment as well as interacting with professionals and policy makers, thus participating in the kinds of decision that will guide the future development of the region. Most aspects of this process are accomplished digitally, through multifarious networks and software which are embedded in everything from the physical pavement on which highway traffic moves to the online transactions which dominate the way the population demand and receive services. The biggest challenge for such an authority is to ensure that such digital infrastructure functions effectively. In such a world, the decisions about what policies to pursue rest with ensuring that the current picture is as up to date as possible, and the public at large are able to drive the process towards the generation of effective and equitable urban plans and policies.

Farfetched? Maybe in terms of our ability to organise such a world but not farfetched in terms of the systems that are now available to achieve such co-ordination. Already such systems exist in the financial services sector, in global stock exchanges, in the rapid move to online banking and in the embryonic development of e-commerce. Test tracks are being laid for automated highway systems, satellites are being launched which will sense the environment to a degree of spatial precision hitherto undreamed of, cities are being hardwired at the local level with closed-circuit TV, and our spending habits are being better understood from the vast array of data that is being synthesised by the many geodemographic companies which have set up business to advise on where best to locate everything which has an economic function. Fifty percent of US households now use the world wide web daily and the number of web pages accessible from the internet now numbers 20 percent of the world's population. The online, digital world fantasised about by the science fiction writers of the near future such as William Gibson and Neil Stephenson, is almost upon us, bringing with it dramatic opportunities for integrating activities that hitherto had been accomplished quite separately, if not conceptualised quite differently as well.

How do we assemble and learn about the nuts and bolts that will make this kind of world possible? And more importantly how can we begin to organise their assembly? This is the task that Robert Laurini sets for himself in this book, the first of its kind. All paths to the future are littered with the debris of attempts that have failed through the inability of their adherents to develop a strong enough organisational model which enables them to integrate many elements that have been hitherto unconnected. The great strength of Laurini's approach is his recognition that at the heart of the digital planning systems that will dominate our future, lie information systems that are being rapidly developed for many different but related purposes. Recognising that good and effective planning systems need to be built around such information is a particular feature of Laurini's approach and one that is essential if urban planners are to learn how to use such systems effectively. Any professional or academic currently confronting the emergent digital world and asking questions about how to use such media in urban planning and policy-making is confronted with many disparate themes which are all presented independently and with little reference to one another. How digital media can be used to visualise everything from land use plans to simulations to statistical analyses to computer-aided designs, how two-dimensional spatial data can be stored, retrieved, analysed and mapped in geographic information systems, how virtual reality systems can be used to provide virtual environments in which users and designers can interact, and how the internet can be used to interact with planning information from the most local of planning applications to the most strategic of policy issues are questions that are usually studied in a vacuum from one another, although they are all part of the digital planning environment that will dominate our future.

What Laurini does in this book is to provide a backcloth for learning about this planning environment and for integrating the many different themes that are beginning to define it. The digital world is about convergence. It is about representing information in digital form in such a way that data and information that previously were considered to be quite unlike one another can be juxtaposed to make their correspondence and linkage considerably clearer. Words and numbers and pictures are of course the critical elements in this new world but as Laurini implies here, sound and touch and taste all have a place in the tools which, he argues, will define digital planning in the near future. This 'convergence of media' offers new insights into the way we might integrate the elements of urban planning. Traditional dichotomies exist between those whose concern is understanding the city – the urban analysts and those whose concern has been to change it – the urban designers and visionaries. Space has been separated from time as designers and policy-makers have sought instant solutions to long-standing urban problems and urban planning has always grappled with change and evolution in a painful and tortuous manner. Scale has separated and dismembered the urban planning environment in undesirable ways as reflected in ways of looking at cities in geographic in contrast to geometric terms, through socio-economic perspectives in contrast to physical perspectives. This separation in terms of scale has been paralleled by a schism between treating planning in abstract terms in contrast to the more down-to-earth concrete terms that are defined by people and buildings.

These are schisms that can be breached by the convergence that the transition to a digital society enables, and it is information that resides at its heart. But it is not just about representing information, for information must be communicated as well as stored and presented. To this end, the information society into which future planning will be embedded, must be based on effective digital communication. As data can be updated in real time through the rapid development of digital monitoring devices and the ability to synthesise data from disparate sources, then the separation of the routine from the strategic is narrowing just as is the separation of events in real-time from those that are defined in a more abstract decision-oriented time. In the past, data which was used for routine management was rarely linked to more strategic decision-making, the recording interval being the important distinguishing issue in terms of data collection. This has now largely disappeared as databases based on routine and often real-time recording can now be used to provide information about longer-term trends and changes. Moreover, the ability of such information to be passed to a variety of rather different users also enables those users to communicate in ways that previously were either difficult and cumbersome, or not possible or even logical. The traditional model of planning which is introduced by Laurini in his opening chapter is based on the notion that problems are analysed and plans developed rationally but usually based on data that was collected or at least organised for such purposes periodically, at much the same rate as the cycle of plan-making. The theory of Western planning was thence based on the notion that a more routine kind of planning management would support these wider strategic processes through frequent monitoring of change, and approval or otherwise of proposals for change which were linked to the goals of the plan. This process would 'keep the plan on course' although in practice the process was beset by *ad hoc* difficulties ranging from the difficulty of organising systems that worked on such different time scales, to the inability of planners to collect and organise coherent information about the current condition. The role of communicating expert information and advice to policy-makers was cumbersome although workable within limits, but the communication of both information and plans to other interest groups and the wider public was usually an organisational failure, despite the best of intentions.

It is the convergence of all these elements through digital media that promises to provide a much more effective urban planning system in the future, one in which the routine and strategic are more closely linked at all levels and one which is informed by consistent real-time databases which draw on new ways of sensing the urban environment. What Laurini does in this book is to demonstrate this convergence in countless ways. His early sketch of the traditional and conventional model of planning is quickly supplemented by ideas about how such information can be communicated between different parts of the planning task. He provides the reader with a thorough outline of database technology with its emphasis on relational and logical linkages. These lie at the core of building digital planning environments in which information can be communicated using the entire range of current interactive technologies from hyperlinking across time and space and within databases to visualising and communicating ideas in virtual environments in which many different interests might be present. He manages to translate these foundations of information into ways in which data can be used intelligently through expert systems, and he illustrates how such systems might form the basis for a new kind of digital planning which is truly integrative and truly collaborative. The elements of such an approach lie in the use of multimedia in interactive working across computer networks and within and between remote databases. The kind of online public participation which he develops is only possible when planning information is available in real time, across networks and through truly visual media.

This model of the future is quite consistent with the kind of grass-roots planning often labelled argumentative or collaborative by theorists such as Healey, Forester, and Alexander amongst others. It reflects the enormous possibilities for integration of diverse and often conflicting viewpoints in the postmodern, postindustrial age. What Laurini does in this book is not only provide us with a road map to the future but also show us how we can build the roads at the same time. In this sense his book is a classic example of how the city itself is becoming digital of which urban planning is just one part and the convergence that he demonstrates will provide the dominant paradigm for planning in the early 21st century. Readers will enjoy what he has to say and come away from his writing much the wiser, armed with knowledge about the elements of these new technologies and the organisational tools to implement them.

Michael Batty
Professor of Spatial Analysis and Planning
University College London
WC1E 6BT, UK

PREFACE

During the last decades, in urban planning, software products were overall used for mapping, storing information and, in a lesser extent for analysis. And now, facing the evolution of computer techniques, conventional urban planning can be totally reconceived.

What was the goal of urban planning? The classical answer was to help drive the evolution of the city, for instance by actions on land regulations. By selecting some key elements in the regulations, a sort of genetic code for the city was designed for conducting the city towards some goals stated by local councillors. Urban planners were seen as technicians fixing the objectives, analysing the possible short-term and long-term consequences, and finally proposing plans in the form of maps and written statements. But the technical limitations have implied not only some organisational limitations, but also limitations in the objectives. By using new technological tools, things which were difficult to control in the past, can today be controlled. Let us think about the place of urban environment and disaster management.

The scope of this book will be to give the readers – urban planners, students in urban planning, even computer scientists – a fresh approach. This book is not 'yet another GIS book', but is rather intended to renew urban planning based on computers. This is not a novel version of big brother but totally the contrary. Computers can help city-dwellers to extend their freedom, to impose their views to urban planners, to share their opinions about the future. In other words, computers must really be liberating tools, not freedom-killing tools. (Refer to the chapters on groupware or on public participation.)

All those new applications must be based on sound data so an information system is the kernel of those novel tools. Older urban information systems were implemented on mainframes without any link to the exterior. Due to new communication techniques, and especially the World Wide Web, new classes of information systems are emerging. More, in 1993 during the EGIS conference,[1] in Genoa, Italy, as an invited lecturer speaking about the future of

---

1   Laurini R., *Updating and Sharing Geographic Information: GIS Challenges for the Year 2000*. Proceedings of the 4th EGIS, *European Conference*, Genoa, Italy, March 29–April 1, 1993. Key-note address, pp. 1651–67.

GIS, I made a prophecy that 'all isolated GIS will disappear' especially because of new communication and telecommunication techniques. Now, web-based GIS exist and interoperable GIS and telegeoprocessing information systems are accessible targets, all characterised by connections with outer systems.

In the title *Information systems for urban planning, a hypermedia co-opearative approach*, I emphasise hypermedia and co-operation. Indeed, not only multimedia data are interesting, but with the multiplicity of links, multimedia information is transformed into hypermedia; see overall the hypermap chapter, and to a minor extent Chapter 7. Secondly, co-operation between machines is another new dimension regrouping aspects such as groupware, interoperability and telegeoprocessing; see Chapters 8 to 11; co-operation between people can be enriched with new tools for public participation in urban affairs.

This book is organised as follows. The first chapter will set the role of urban planning within a context of systems analysis, so preparing for a theoretical basis of novel computer tools such as co-operative work. The second chapter will give some elements for the design of information systems emphasising the multimedia object-oriented aspects. Data acquisition in urban context will be the core of the third chapter passing gradually from surveying to GPS and sensing. Data quality control and multisource updating will constitute the fourth chapter. The fifth chapter will be devoted to hyperdocuments and especially hypermaps in order to design web sites for urban planning, possibly under the form of intranets or extranets. Everything regarding knowledge engineering and metadata will be regrouped in the sixth chapter. Then the various techniques of visualisation will be presented in the seventh chapter, not only for displaying the database contents, but also new techniques for accessing and navigating within the information space. Groupware and computer supported co-operative works for urban planners will be discussed in Chapter 8. Since public participation is a key element of the success of urban planning, Chapter 9 will emphasise devoted techniques such as virtual reality and web-based information sharing between citizens. All problems regarding computer architectures will be regrouped in the tenth chapter. The last chapter will present novel tools with wireless telecommunications for monitoring the urban environment and managing urban risks and disasters.

I have to acknowledge several persons. First of all, I have to thank various researchers of my lab, or people who have worked with me in the past: they all allowed me not only to pioneer several fresh research directions but also to develop and implement new tools, Françoise Milleret-Raffort, Sylvie Servigne, Marie-Aude Aufaure, Thierry Ubeda, Bruno Tellez, Christine Bonhomme, Tullio Tanzi, Azedine Boulmakoul, Franco Vico, Alain Puricelli, Jean-Claude Müller, Luigi Di Prinzio, Ahmed Lbath to name a few. I have also to thank all the members of my family for the time I have spent writing chapters, making drawings, browsing the web, participating to conferences, instead of being at home with them.

When writing this book, it was more and more difficult for me to write with a conventional style. For each new paragraph, I had to include some WWW references, either for multimedia or for co-operative aspects. Facing this difficulty, one possibility would have been to include a CD-ROM together with this book; but it was decided to create a web site with links to interesting Internet pages. Visit the site http://lisi.insa-lyon.fr/~laurini/isup/, in order to include new examples and references. All interested readers can send messages and references in order to enrich this site.

Professor Robert Laurini
Villeurbanne
http://lisi.insa-lyon.fr/~laurini
Email: laurini@if.insa-lyon.fr

1

SYSTEMS ANALYSIS OF THE URBAN PLANNING PROCESSES

The activity of controlling a city or a territory is very important because their evolution is not only determined by heavy trends, but also in a context of co-operation and conflicts between several actors. In this perspective, an urban information system, or more exactly an information system for urban planning (ISUP) must be the ideal tool allowing some kind of decision or negotiation support between several actors. Thanks to systems analysis and the theory of the general system,[1] a very relevant approach will be offered giving a framework for analysing the city and providing tools for solving existing urban problems by means of action plans. The main difficulty is not only to implement urban plans in order to solve those problems, but also to monitor very carefully the evolution of city-wide activities and phenomena with the assistance of information systems. Moreover, computers will be used also for daily urban management in order to reach the objectives listed in the plan.

In this chapter, the city will first be examined from the viewpoint of general systems, so giving some theoretical background for finding or implementing solutions to urban problems. Within this framework, decisions will be examined, and especially computer-based decision-supported systems in order to emphasise the importance of plan-making, and, in a more general perspective, the differences between urban planning and urban management especially with the target of sustainable cities. Some discussions about the role of information systems for urban planning especially as a co-operative process will also be given in order to conclude this chapter.

1.1 Modelling cities as general systems

A city can be described as a general system (Laurini 1978, 1979, 1980) with different interconnected subsystems. According to von Bertallanfy (1968), a

---

1 The conventional expression 'General System Theory' is ambiguous, because some distinctions must be made between the general theory of systems and the theory of the general systems. In this chapter, we will use only the latter meaning.

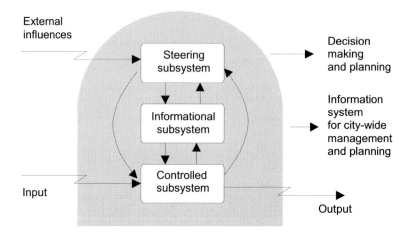

Figure 1.1 A city as a general system

system is *a complex of objects standing in interactions*. But Lemoigne (1977) has studied all organisations, and especially human organisations as systems presenting very special characteristics, known as general systems which are characterised by function, evolution, environment, finality and structure. He said 'a general system is an object doing something within an environment, provided with a permanent structure, able to evolve and generally being given some finality'. So it looks possible to model a city as a general system including (Figure 1.1):

• a steering subsystem
• an informational subsystem
• a controlled subsystem.

Let us examine all of them, knowing that some limited recursiveness can apply. Indeed, as an example, within the informational subsystem a sort of steering system can be identified, and so on.

1.1.1   The steering subsystem

The role of the steering subsystem is to design and make decisions in order to shape the global system and make it evolve towards the desired directions. The main components are (Figure 1.2):

• determination of objectives
• diagnosis of the problems
• design and selection of decision alternatives
• planning.

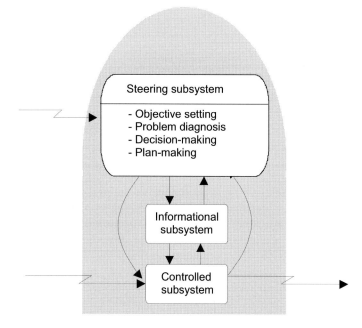

Figure 1.2 Components of the steering subsystem.

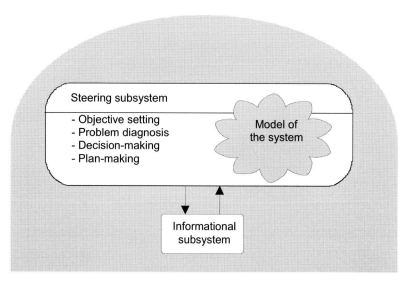

Figure 1.3 Any steering subsystem has a model of the complete system.

By its nature, this subsystem can be considered as multi-actor, because a great number of persons, associations, companies and administrations are trying to in-fluence the evolution of the city. Those implications will be studied later in this book.

Another important consequence is that any steering system must integrate a model (Figure 1.3) of the complete system; here we mean not only a model of the controlled system, but also a model of the information system and of the steering system. Sometimes the expression 'mental model' of the system is used. In our case, we will speak of a mental model of the city that each actor has in his mind.

1.1.2 The controlled subsystem

This subsystem includes all elements for which the decision will be made: it regroups all physical sectors and sociological phenomena which can be influenced by decisions. Among sectors, let us mention demography, employment, housing, land use, public services, budget, environment, transportation and so on. Those components can interact as illustrated in Figure 1.4.

Since 'everything is connected to everything', it is very important to emphasise that any modification can have consequences on the other components. So, when a decision is made on a particular sector, implications can occur or will occur in

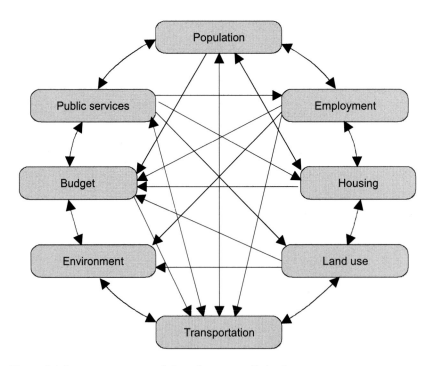

Figure 1.4 Some components of the urban controlled subsystem.

other sectors. In other words, if the goal of some planning activities is to make some variables evolve to some desired targets, one must take account of the possible repercussions on the other sectors. In other words, when solving an urban problem, it is important not to create voluntarily or involuntarily the conditions or the background for newer problems, possibly with consequences much more drastic and more difficult to solve.

1.1.3   The informational subsystem

The role of the informational subsystem is to regroup all information necessary for controlling and steering a city. Generally speaking there can also exist some non-computerised informational systems, but more and more computerised information is the cornerstone of the information system. More precisely, an information system for urban planning must integrate all possible strategic information coming from the steering subsystem, and all measures made on the controlled subsystem. In the past, this subsystem was not identified very well because information was not formally included in an identified subsystem, perhaps within some paper-based systems. Nowadays, this subsystem is materialised by some computer systems as databases.[2]

For correct functioning, this information must be free from errors, regularly updated, etc. In other words the stored information must represent an high-fidelity model of the city, and when data are coming from different sources, the results of the integration must be consistent in connection with the objectives (Figure 1.5).

Strategic information

Information coming from the steering subsystem essentially concerns the objectives, alternatives, and criteria for comparing the alternatives and their evaluation, etc. Experiences show that it is very difficult to integrate all those pieces of information especially because of the multiplicity of actors who can have very divergent interests. For instance, if an information system belongs to the municipality, then it will reflect primarily the interests of some city council members and to some extent the interests of some other decision-making bodies (regional level for example) and of some particular urban groups. More generally an information system's contents reflect correctly the interests and the objectives of its owner; there is always a gap in representing interests and objectives of the opponents whoever they are.

---

2   By database systems, we mean here the computer subset of an information system. Differently said, an information system can include some other aspects which are not computerised, such as fax systems.

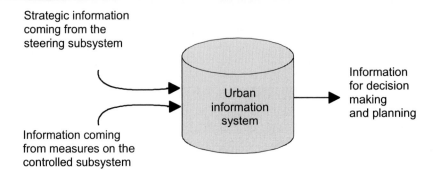

Figure 1.5 Main sources of an urban information system.

Sometimes, some local authorities support some limitation in land-use planning due to the existence of a master plan imposed by some upper-level administration. If needed, those aspects must be included into the strategic information. As incoming information is relatively difficult to organise, experience shows that outgoing information is easier to model by using spreadsheets, administrative forms, diagrams, maps and so on.

Information coming from the controlled subsystem

This information is relative to the sectors as illustrated in Figure 1.4. Of course, other sectors can be concerned such as cultural activities and tourism. In general, acquisition of this kind of information is considered easier, however, a huge handicap can be the volume. For a big city, billions of objects must be described (parcels, persons, buildings, pipes, trees, etc.) leading to zillions of data to manage (factual information, shape, positions, temporal evolution and so on). For years, stress was placed on alphanumeric data coming especially from administrative forms. But more and more other types of information are considered, visual (not only maps but also photos or images) and auditory information (voice, urban noise and any kinds of signals). This multimedia information is now more and more common in urban planning. Chapter 3 will address those issues.

1.1.4   On the controlled domain

In theory, decisions have or must have consequences on the territory to be managed. But in reality, when a mayor[3] decides to open a new industrial

---

3   In this book, I will denote by mayor, the strong man in the city council (or in any local authority), that is to say the person who has effectively the power to decide. I am aware that in some countries, the power is devoted to the president of the local council, whereas the lord mayor has only a representative role.

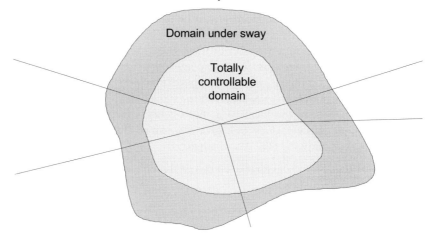

Figure 1.6 The totally controllable domain, the domain under sway and the domain outside any influence.

zone, employees not only come from his city, but possibly from neighbouring cities. Even if the mayor's initial idea was to reduce unemployment in his city, employment in his own city will be reduced less than anticipated, because some newly employed people will also come from neighbouring cities, and so the reduction can apply to the cities located in the vicinity. This small example illustrates what can be distinguished between three domains (Figure 1.6):

- the domains which are totally under control of the steering subsystems (totally controllable domains)
- the domain which is partially controllable and from which some influence must be possible (domain under sway)
- the domains which are totally outside any influence.

Obviously, conversely, the domain (or territory) the mayor wants to control is also in some part a domain under sway of some other organisations. In other words, their actions will influence his domain more or less. In reality, the problem is much more complex. Theoretically, no sector is totally controllable and there is no domain outside of any influence. In other words, a sort of continuum of controllability can be defined. Perhaps the theory of fuzzy sets can be applied to this issue (Negoita and Ralescu 1975).

1.2 Urban problem solving

According to Hayes (1981) a problem can be defined as follows: 'Whenever there is a gap between where you are now and where you want to be, and you don't know how to find a way across that gap, you have a problem'. He continues by defining solving as 'finding an appropriate way to cross that gap'.

What is an urban problem, what are its particularities? The first theoretical answer is to declare that any urban problem (Castells 1977) does not exist by itself because it is only the facet of a more general societal problem. This debate-expanding answer does not give any more information about its nature. In 1971, Chadwick (1971) was saying:

Problem = goal + impediment to the goal

This definition was extended by Faludi (1973) declaring that 'a problem is a state of tension between the ends pursued by a subject[4] and his image of the environment'. So, three aspects can be pointed out:

- the importance of the actor (person, group, organisation)
- the perception of reality
- the possible future.

We then see a little more clearly what urban problems are, together with a sort of disease because the great difficulty about urban problems is their identification. According to several urban actors it is not worthwhile to develop sophisticated instruments to identify them; some major actors prefer waiting to see the problem set by some other actors. For instance, let us examine very rapidly the noise pollution in cities. If you look at claims sent by residents to the city hall, it is very interesting to see that:

- no complaints are sent from people living in the vicinity of very noisy motorways
- the majority of complaints come from very quiet residential areas.

If a city planner were to develop a strategy to deal with urban noise by using only the number of complaints received, the results would be very strange; he would plan to reduce noise in very quiet areas and do nothing in very noisy areas. However, more or less, all urban problems are linked together to form a sort of mess. An example is pictured in Figure 1.7.

The main issue is to identify solutions. Urban problem solving is a very difficult challenge, perhaps the most difficult our society has presently to face. Alas, very often, by trying to solve a single problem, some new ones can emerge. For instance, in the 1960s in France, the housing problem had to be solved so a broad policy of housing was launched especially with big and

---

4   Now, the term 'actor' is preferred.

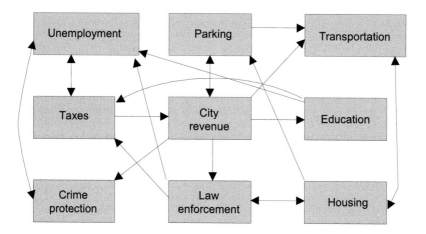

Figure 1.7 A mess of urban problems.

tall buildings. Some years later, delinquency and violence emerged in these buildings due to overcrowding. This example emphasises that a decided action set A_1 for solving a class of problems P_1 yields new problems P_2 needing some action set A_2 and so on. A sort of problem-action chain is emerging, or more exactly a loop as illustrated in Figure 1.8a, in which 'increasing problems' can be read as 'increasing the importance of the same problem' or 'creating new problems'.

If one can say 'the more actions we implement, the more problems we get', this is a vicious circle (1.8b), sometimes also called a snowball effect. When it can be said 'the more actions, the less problems', this is a virtuous circle (1.8c). Indeed, experiences have shown that in very particular circumstances, actions implemented for solving an urban problem can help to solve others or to

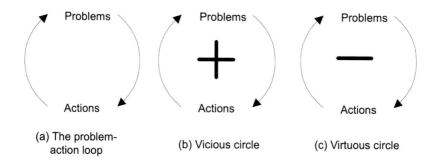

(a) The problem-action loop

(b) Vicious circle

(c) Virtuous circle

Figure 1.8 The problem-action loop and different cases. (a) the general loop. (b) the vicious circle. (c) the virtuous circle.

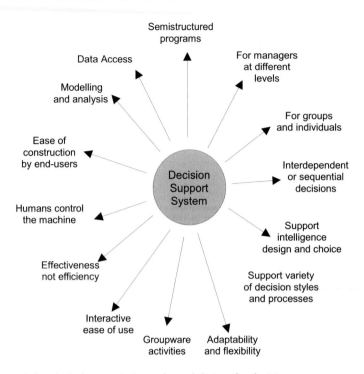

Figure 1.9 The ideal characteristics and capabilities of a decision-support system. According to Turban and Aronson 1998.

diminish their acuity. So, one of the goals of urban planning will be to find actions able to solve several problems either directly or later.

1.3 Decision-support systems

Decision-support systems, often known as DSS in the literature are computer systems able to assist the decision-makers by analysing issues and proposing solutions. More precisely, for Turban and Aronson (1998), a decision-support system as idealised in Figure 1.9 must support:

- decision-makers for all kinds of real-life problems, especially for semistructured or unstructured situations
- different kinds of managerial levels
- individuals as well as groups; in this case, we mean co-operative systems
- sequential or interdependent decisions
- all phases of the decision-making process: intelligence, design, choice, implementation
- a variety of decision-making processes and styles
- flexibility and adaptability over time

- friendliness for all kinds of users
- effectiveness (accuracy, timeliness, quality) rather than efficiency
- easy construction of new models
- modelling and analysing problems
- accessing all kinds of data.

In addition, it must provide some groupware facilities.

So a decision-support system can be defined as a computer system helping one or more actors in their work of making decisions. One of the key components consists of what-if models to simulate the future of the city according to some assumptions. Among those urban models, let us mention:

- transportation and traffic models
- pollution models
- service and commercial premises location
- energy and water consumption
- water production
- etc.

Figure 1.10 illustrates the main components of such a system which includes:

- **acquisition of strategic information**, that is to say information coming from the steering subsystem, together with acquisition of information about the territory under control;

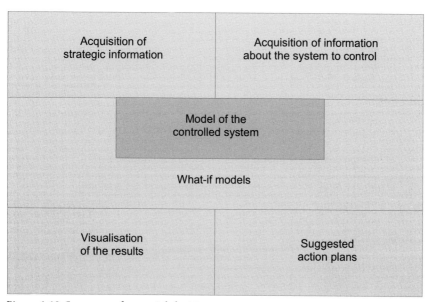

Figure 1.10 Structure of a spatial decision-support system.

- **acquisition of information about the system to control,** that is to say information coming from the controlled system by means of any kind of acquisition techniques or measuring instruments;
- **a model of the controlled system** in order to project or forecast evolution; by projection, we mean that it is the continuation of the past, all things being equal; if some parameters or assumptions are changed, one speaks about alternatives and forecasting;
- **modules of what-if models for data analysis and system simulation;** the role of this component is to study the data of the past in order to find some regularities for constructing the model; when some performance indices exist, it is possible to evaluate and compare the effects of the simulated alternatives;
- **visualisation of the results;** for any alternative, the main variables can be displayed in order to compare them visually;
- **action plans,** i.e. when an alternative is selected, some action plans must be drawn up and implemented.

Two kinds of decision-support systems can be proposed, (a) based on operations research or (b) based on document navigation as given in Figure 1.11. Those which were built on urban models (Foot 1981, Sui 1998) can be

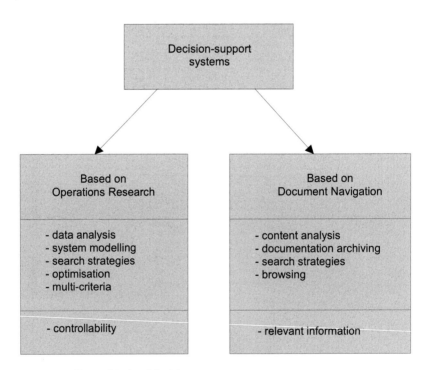

Figure 1.11 Different kinds of decision-support systems.

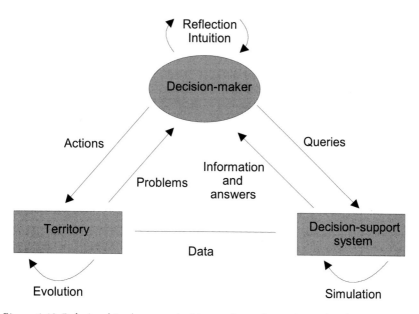

Figure 1.12 Relationships between decision-makers, the territory they have to control and the DSS.

considered as based on operation research, whereas those based on document navigation will be presented in Chapter 5.

In our case, three main elements interact, the decision-makers, possibly in conflict with each other, the territory they have to control and the decision-support system itself (Figure 1.12). By means of simulations of the city, the decision-makers can see how the city evolves due to their actions. When the simulated evolution looks satisfactory the action plan is likely to be found.

1.4   Urban plan-making

The design and implementation of a plan is something very delicate, not only concerning its content, but also in terms of project management and of tasks to be performed. In this section, several views of processes will be examined.

1.4.1   Definition of urban planning

According to Doubriere (1979) urban planning must make the city healthier, bigger and nicer, together with safeguarding and showing to advantage the city's heritage. But Henderson (1997) is convinced that no single definition can possibly cover the breadth of information and practice that makes up contemporary planning. For him, urban planning is best thought of as a process that uses a variety of tools (zoning, transportation planning, environmental policy, housing programs, etc.) to achieve envisioned and desired goals within the natural and built environments. There are varieties of professional activities that may be considered planning, including land-use planning, transportation planning, environmental planning, social planning and urban design. Each of these has different underlying theories, technical practices, and prominent leaders. However, most planning activities have the following four qualities in common:

1. **Planning is future-oriented.** The decisions made in the planning process are generally made to affect a future condition in the environment. This seems obvious, but it is a good thing to reinforce (see Section 1.5.2).
2. Planning is concerned with defining and **evaluating alternative solutions** to problems. This is deeply rooted in rational planning theories that underlie current planning practice. It is based on the premise that it is difficult to argue that a chosen strategy is the right one to pursue if alternatives have not been defined for comparison and evaluation.
3. **Planning is political.** Every public planning decision takes place in a political context. It is important to realise early on that the majority of planning activities involve the use or regulation of land in some form, and that all land belongs to someone who is afforded rights, most importantly the right to due compensation.
4. **Planning has a special** responsibility to represent the needs of minorities, the disabled, the poor and other underrepresented groups. Planners are required to pay special attention to the needs of these groups as part of their professional code of ethics.

For other definitions of urban planning, see the New Charter of Athens.[5]

1.4.2 Legislative framework

As an example, the French land-use plan-making procedure will be examined in this section. The main French regulations are integrated in the POS which means land-use planning. Each local authority must design this kind of plan whose goals are:

- organising urban areas
- protecting natural areas
- preparing for the provision of future public infrastructures and social equipment
- providing a legal context for any new construction.

More precisely, this plan:

---

5   http://www.byplanlab.dk/english/athen.htm or http://www.ceu-ectp.org/en/athensen.htm

- delimits the urbanisation zones taking account of soil characteristics
- determines the assignment of zones according to planned occupation (industrial areas, rural areas, etc.)
- determines for each zone the floorspace ratio which is the ratio between total floorspace of the plot and the area of the plot (this ratio provides a simple way to measure building density)
- determines new roads and their modification
- gives the list of protected areas (architectural, archaeological, forest, etc.)
- etc.

A POS must include map documents and written documents. For each zone, 15 articles are given as follows:

Article 1: Forbidden land use types
Article 2: Land use types submitted to special conditions
Article 3: Road and accesses to constructions
Article 4: Engineering networks (water, electricity, sewerage, etc.)
Article 5: Plot characteristics (shape, area) for accepting buildings
Article 6: Location of buildings *vis-à-vis* roads and public spaces
Article 7: Location of buildings *vis-à-vis* neighbouring limits
Article 8: Location of buildings within the same plot
Article 9: Building footprint to plot ratio
Article 10: Heights
Article 11: External appearance of buildings (shapes, materials, colours, etc.)
Article 12: Parking and parking lots
Article 13: Green spaces
Article 14: Floor space ratio
Article 15: Possibilities of exceeding the maximum floor space ratio.

By applying these rules, a sort of building envelope can be defined as given in Figure 1.13 and an example of a map statement is given in Figure 1.14.

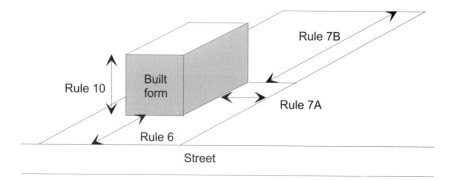

Figure 1.13 Applying building rules to define a building envelope.

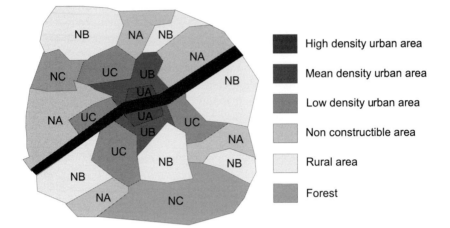

Figure 1.14 Example of French land-use map (POS).

As a second example, let us examine the procedures in the UK for development planning (Figure 1.15) (Rydin 1998). Once the plan is made at a technical level, it is deposited. Normally a Public Local Inquiry (PLI) or an Examination in Public (EIP) is organised in case of objections. Then a report is written. Finally some modifications can be accepted. Ultimately the British Secretary of State adopts it when he is satisfied about any modifications.

1.4.3 Systemic view

In Figure 1.15, the legislative steps of a plan in the UK are presented as an example starting from a draft of a plan. But the main question is how to make a plan at strategic, tactical and operational levels? The first step is to start from the concepts of systems analysis in order to draw up a list of all phases to be carried out. Such an example is given in Figure 1.16 and taken from the systemic view (Sarly 1972) with some minor modifications. The first step is the 'recognition and the definition of the urban problems' and the last one 'monitoring and review'. One of the key steps is 3 Data collection, that is to say the acquisition of all data necessary to foster models for projections. Another key step is 7, the design of alternative plans. The main assumption behind this procedure is that there is some sort of consensus, and the existence of no conflicts between actors.

1.4.4 Conflict-tackling view

Another view of the plan-making process is to see it as the seeking of a solution between different actors in conflict. Under this assumption, a solution

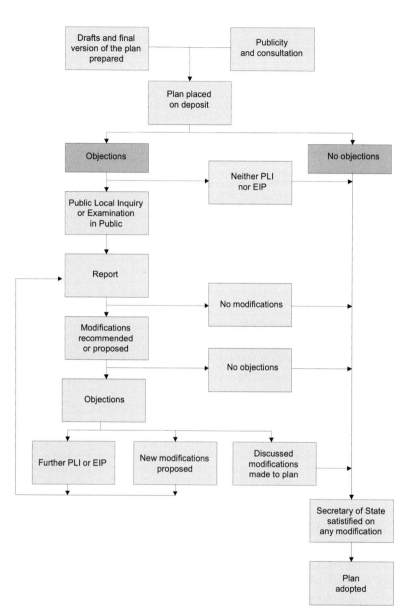

Figure 1.15 The development plan process in the UK, according to Rydin (1998). Adapted from Planning Policy Guidance note 12 section 4. PLI means Public Local Inquiry and EIP Examination in Public.

can be found as pictured in Figure 1.16, in which the starting point is the list of requests presented by different politicians (1), the discussion with them (5), the arbitration (10) and the proposal of a compromise for spatial organisation and service location (11). In this case, the procedure of plan making is seen as the search for a consensus between some conflicting demands. But when the conflict is very fierce, could we imagine what the consensus would be between two prize-fighters?

1.4.5 Participative view

Another example is taken from the structuring presented by a French politician very keen on people participation; the whole process is based on the participation with resident associations as illustrated in Figure 1.18. This process can be seen as a variant of conflict resolution, but with a different spirit; the main actors are not politicians, but resident associations.

1.4.6 From systematic to co-operative plan making

Figures 1.16–1.18 illustrate not only different views of the planning processes, but present overall different philosophies. The first can be seen as a model of a unique decision-maker needing some consultants (Figure 1.16) to help him (monarchy or despotism), the second (Figure 1.17) in which several politicians are involved (oligarchy) and the last (Figure 1.18) based on participation (democracy).

More and more nowadays, plan-making procedures involve public participation (participative information systems), whereas co-operation will mean that several actors (urban planners, politicians, resident association spokesmen, etc.) working together for the objectives, and therefore for the same action plans. In this book, emphasis will be given overall on co-operation (see Chapter 8) and public participation (see Chapter 9).

1.5 Urban planning versus urban management of sustainable cities

As the main role of urban planning is to design new plans taking some objectives into account, which can be seen as the long-term part of the plan-making process, urban management can be seen as the daily process in order to follow and monitor the plans. In this section, we will try to make some comparisons between urban planning and urban management. After that, some elements for planning and management of sustainable cities will be given in order to emphasise the connections between short-term and long-term consequences.

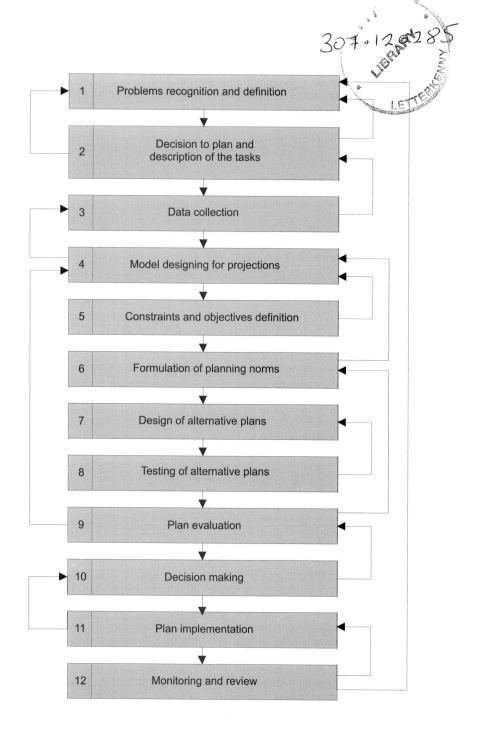

Figure 1.16 A systematic view of the urban planning process.
Simplified, according to Sarly 1972.

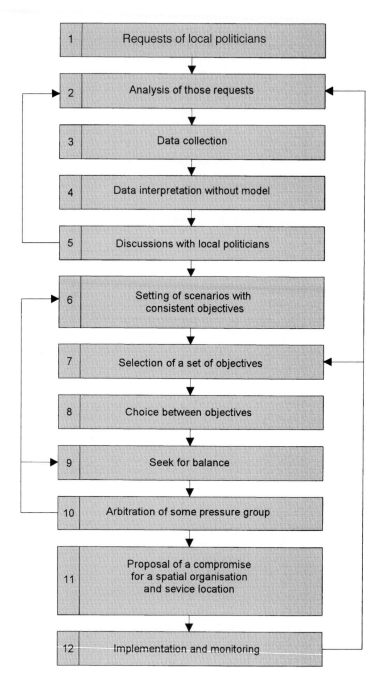

Figure 1.17 Plan-making process based on conflict resolution.

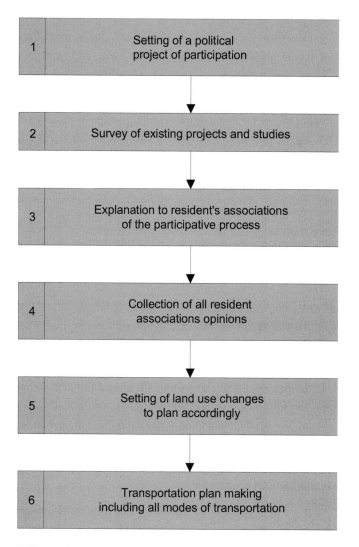

Figure 1.18 Plan-making process of a local French politician backed by resident associations.

1.5.1 Urban management

Urban management means:

- data collected on a daily basis to foster municipal information systems (accounting, building permits, etc.)
- different kinds of administrative authorisations
- maintenance of the city infrastructures

- social services
- etc.

Figure 1.19 illustrates the comparison between an urban planning system and a monitoring system. As input, the planning system receives some objectives for the future, and by means of information on the past and the present of the city, a plan for the future will be designed. In addition, this plan for the future can be seen as the input of the monitoring system; so the state variables of the city as produced in the monitoring system, can be easily compared with the references as given in the plan for the future.

For an elected official, urban planning reflects a mid-term or a long-term vision of the city, whereas the urban monitoring or management system is for the short term. One of the main difficulties in urban planning is that the length of the elected officials' mandates is very short, for instance, five years and for far-sighted actions maybe 20 years. This time difference reveals the difficulty some elected officials have being comfortable with long-term urban planning.

Even though the subject of this book is not urban management, there are many connections between urban planning and urban management. Among others, urban planning assumes and implies that daily management is correctly done; and if not an objective of urban planning can be to correct this. To conclude this very short section, let us say that urban planning is more concerned with the long term whereas urban management is more concerned with the short term.

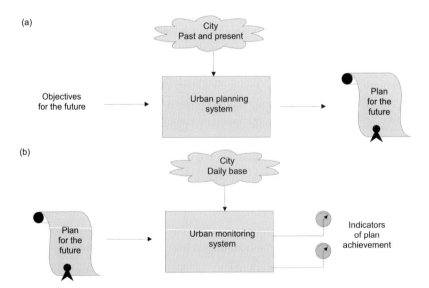

Figure 1.19 Comparing systems for (a) urban planning. (b) urban monitoring.

1.5.2   Planning for sustainable cities

One of the main concerns of urban planning is the concept of sustainability. Sustainability is defined by the World Commission on Environment and Development (WCED 1987) as the 'development that meets the needs of present generations without compromising the ability of future generations to meet their needs and aspirations'. According to May *et al.* (1996), the fundamental principles of sustainable development are as follows:

- **Futurity** (also known as inter-generational equity or 'not stealing from our children and grandchildren')
- **Social equity** (also known as intra-generational equity or 'care for today's poor and disadvantaged')
- **Public participation**, in other words, individuals should have an opportunity to participate in decisions that affect them and in the process of sustainable development
- **Environment** (environmental conservation and protection).

The scope of sustainability is that every citizen, present or future has a decent quality of life. But quality of life must represent a balance or a compromise between a number of issues as depicted in Figure 1.20, considering physical

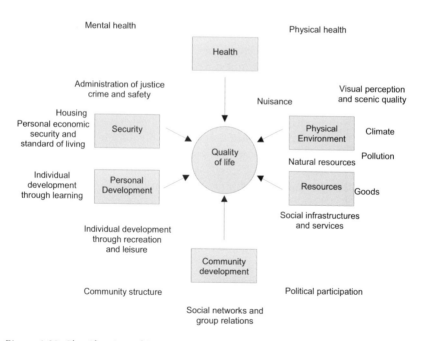

Figure 1.20 Classification of issues concerning quality of life in cities. From May *et al.* 1996.

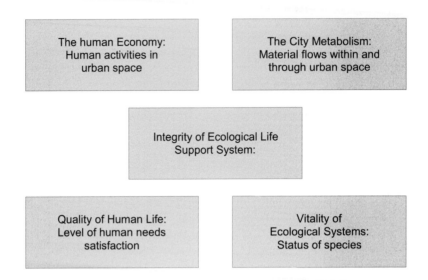

Figure 1.21 Components of a conceptual framework for a sustainable city model. From May *et al.* 1996.

and mental health of persons as well as the situation of the community and public participation. All those issues can be integrated into a model the general framework of which can be seen in Figure 1.21.

In order to reach those objectives, one of the possibilities is to develop a model, or more exactly a set of integrated sub-models, of a sustainable city. According to May *et al.* (1996), any model should always:

- avoid nebulous relationships
- develop sub-models for particular places, where equations can be specified with more confidence
- emphasise problems which are difficult to manage, have complex causes and wide-ranging impacts
- enable a contrasting range of policies to be tested.

To fulfil all these objectives a strategy is proposed based on the following steps:

1 develop **performance indicators** for sustainable cities (i.e. called sustainability indicators)
2 develop a conceptual (i.e. theoretical) model of city sustainability covering all issues of current and potential concerns
3 develop a system architecture capable of mounting as wide a variety of existing operational models as possible
4 assess the suitability of existing operational models for use as building blocks within the conceptual model structure

Table 1.1 Indicators of sustainability (according to May *et al.* 1996)

Theme	Proposed indicators
Resource depletion	Fraction of energy use generated from non-fossil fuel resources. Annual water use as a percentage of the stock in a 50 year return period drought. Percentage of households by key demographic group spending more than 10% of their income on domestic energy supply. Area by regional habitat type.
Residual emission	Respiratory illness in children under 16 years. Greenhouse gas emission. Exceeding critical acid load for sensitive agricultural soil. Remaining capacity at existing landfill sites.
Social sustainability	Life expectancy in years at birth. Number of people in temporary local accommodation. Staff student ratio per year. Total number of violent and non-violent crimes. Number of households at or below the official poverty line
Ecological integrity	Abundance of keystone and flagship species in key regional habitat.

5   mount and link operational models on the model system in a manner consistent with the conceptual model

6   develop operational sub-models to fill modelling gaps evident from the conceptual model structure

7   assess data availability and quality for the operational sub-models, and implement data collection schemes to address data gaps

8   test model predictions against observed data

9   repeat steps 4–8 until model addresses sufficient policy and scenario variables, processes and sustainability indicators, and fulfils objectives identified above.

The proposed sustainability indicators (May *et al.* 1996), are shown in Table 1.1.

1.6   Role of information systems

Describing the role of information systems in urban planning amounts to speaking also about the importance of new technologies in organisations. For centuries information systems were based on paper. Computers and telecommunications present something very different. Let us detail some aspects.

1.6.1   Technology and information systems

To understand the general relationships, it is interesting to base our understanding on three studies:

1   **Leavitt diamond** (Figure 1.22) (Leavitt 1965, Keen 1981). This diagram stresses the relationships between the structure of the organisation, the used technology, the staff and the tasks to be performed in a sort of balance between those components. The juridical framework shapes the whole whereas the output can be seen as the level of services to be provided.

2   **Strategic triangle** from Tardieu and Guthmann (1991) (Figure 1.23). This diagram illustrates the relationships between the structure of the organisation, its global strategy and the level of information technology. Another way to examine this triangle is to see that if there exist two organisations (for instance two different cities) with the same

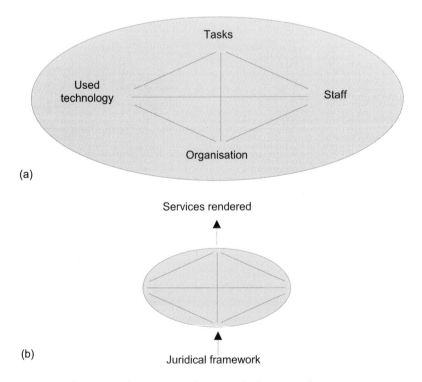

(a)

(b)

Figure 1.22 The Leavitt diamond. (a) the original. (b) revisited.

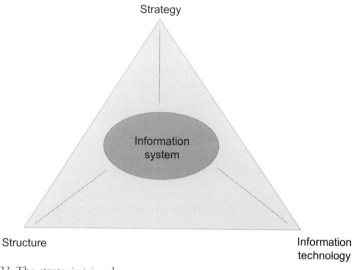

Figure 1.23 The strategic triangle.
According to Tardieu and Guthmann 1991.

structure and the same level of information technology used, the information system will also depend upon the strategy of the decision-makers.

3   **The government pyramid** of Huxhold (1991) (Figure 1.24). The last diagram depicts the relationship between processes, management and policy within organisations. Information comes up from processes, fosters

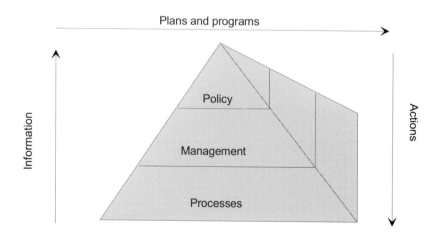
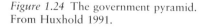

Figure 1.24 The government pyramid.
From Huxhold 1991.

management and is the basis of policy-making, whereas actions go down from policy, management, and finally are implemented within processes. One passes from information to actions by means of plans and programs.

As a conclusion to this section, it is now possible to better understand the links between information technology, strategy, organisation, action plans and so on. Therefore, designing an information system is not reducing it to the design of the database, but integrating all organisational aspects surrounding it.

1.6.2 Group collaboration and information systems

Dealing with changes is one of the most fundamental challenges in the design of information systems. Recently De Michelis et al. (1998) have proposed a new vision of the design by emphasising the three-faceted aspects. For them, the change is governed and arises from three areas of concern – computer systems, group collaboration and organisation – and the interactions among them are illustrated in Figure 1.25.

As the computer systems and the organisational facets are traditionally studied, the third facet will be further examined in this paragraph. So, the group collaboration facet is concerned with people working on a common process, or an *ad hoc* project (for instance co-authoring a document or designing a land-use plan). During such collaborations, people co-ordinate

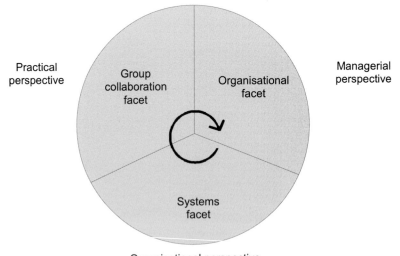

Figure 1.25 The three facets of information systems.
Source: De Michelis et al. 1998.

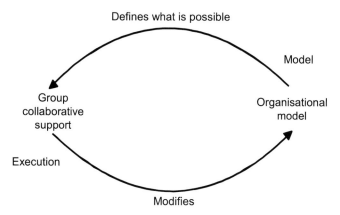

Figure 1.26 Interdependencies between group collaboration and organisational facets.
Source: De Michelis *et al.* 1998.

their activities, deal with contingencies, and change their practices through discussion and learning.

This will be the basis of groupware and of Computer Supported Co-operative Work (CSCW), which will be developed later in this book (Chapter 8).

The rules, roles, and goals characterising the interaction within a work group are defined in the organisational model. Conversely, to resolve problems, group collaborators may modify the organisational model as illustrated in Figure 1.26.

1.6.3 A GIS as a building block of an urban planning information system

Geographic information is a subset of the information system for urban planning. More exactly, it is common to see software products actually named as a GIS (Geographic Information System) to be present in the comprehensive information systems of any local authority. In addition to the geographic database which is the core of a GIS product, Figure 1.27 illustrates four groups of functionalities which are currently seen:

- a subsystem for geographical data acquisition
- a subsystem for spatial analysis
- a subsystem for cartographic presentation
- a subsystem for data management.

Moreover, the geographical database needs to be regularly updated, and sharing information with other GISs is also increasingly important. Those aspects will be studied later (Chapters 4 and 10).

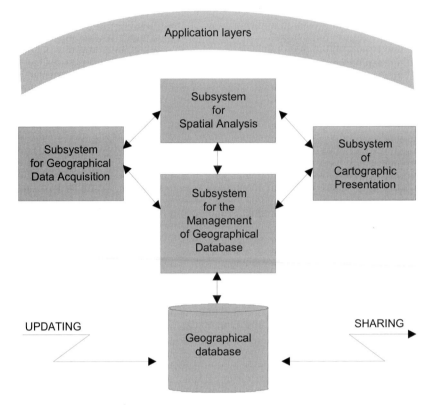

Figure 1.27 Structure of a GIS.

1.6.4 Urban modelling and GISs

As previously explained, it is often important to use mathematical modelling techniques in order to understand the past or simulate the future; they are especially interested in simulating or forecasting strong trends, and to test assumptions (what-if models). An example for modelling the connections between land use and transportation is given in Figure 1.28. Starting from a migration model and the planning regulations, it is possible to make a first forecast land use. Then by connections to transportation models, the final equilibrium can be found between land use and transportation. Please visit the following site for a collection of urban models http://www.odot.state.or.us/tdb/planning/modeling/modeling.html.

It is possible to create different kinds of models:

* **strategic models** are used to support top management strategic planning responsibilities; they must be broad in scope and essentially based on external data

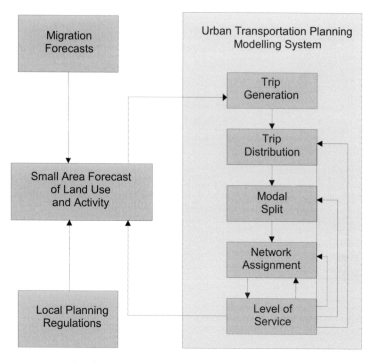

Figure 1.28 Example of an urban planning model for the interactions between land use and transportation.
Source: Heikkila *et al.* 1990.

- **tactical models** are used mainly by middle management to assist them in their tasks
- **operational models** for daily management.

During the last decades, research focusing on urban models was more or less abandoned, especially after Lee's paper in 1973 entitled 'Requiem for large-scale models.' Now, due to the existence of new modelling techniques and the existence of GISs, new generations of urban models will emerge especially based on the paradigm of agents also known as software agents (Bradshaw 1997). This methodology tries to simulate the behaviour of actors by means of rules, instead of equations as in conventional urban modelling techniques. When an organisation is using several models, often a model management system, also called model base, must be implemented the structure of which is given in Figure 1.29.

Several kinds of connections linking a GIS and an urban modelling package can be envisioned as illustrated in Figure 1.30. We can see the case in which the GIS is embedded into the urban modelling package or in contrast the modelling package is embedded into the GIS. Another possibility is to have a loose or a tight coupling between the GIS, the modelling package and the

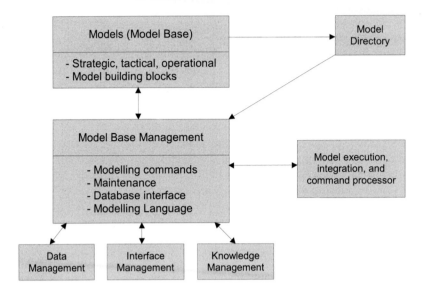

Figure 1.29 The structure of the Model Management System.
According to Turban and Aronson 1998.

statistical package. For technical reasons, the preference is for the tight connections (Figure 1.30d). After having examined urban modelling aspects, let us move on to the role of actors in urban planning.

1.6.5   *Actors and information systems*

An actor can perform many actions and any action can be acted by several actors. Figure 1.31 stresses the many-to-many relationships between actors and actions towards the real world, or more exactly the real city they have to govern. If synergies are very interesting to reinforce, sometimes some opposition can occur between different actors as they have divergent goals.

In Figure 1.32, one can see a group of urban planners supported by an information system trying to elaborate actions in the same city in order to solve urban problems. A very important aspect is their mental model of the city. The city mental model (CMM) is the image of the city that everybody has in mind. This image generally is a particular vision of the city, past, present and future. It helps anyone to understand the city, and each action brings some additional knowledge in this model in order to design a comprehensive information system.

In other words, an information system must integrate those different mental models in order to be helpful. In Chapter 2, under the name of external model, a design methodology will be presented in which each user must give the essential aspects and components of his mental model.

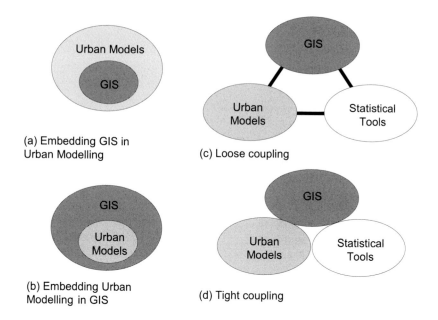

Figure 1.30 Integrating a GIS with an urban modelling package: current practices. (a) embedding a GIS in the modelling package. (b) the reverse. (c) loose coupling. (d) tight coupling.

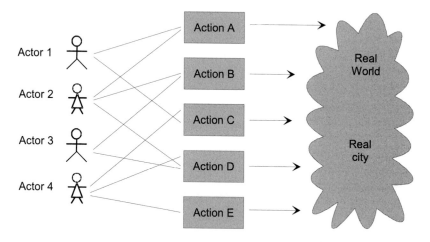

Figure 1.31 Multiple actors, multiple actions towards the real world/real city.

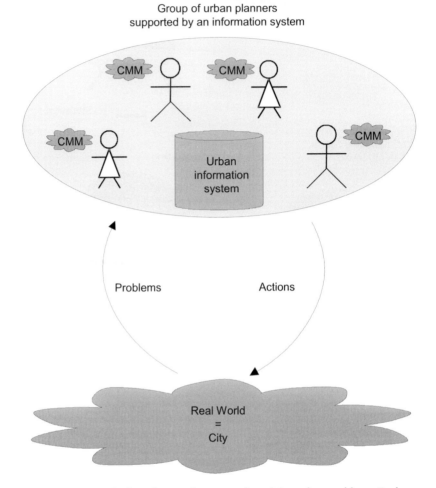

Figure 1.32 Group of urban planners in synergy for solving urban problems. Each actor has his own representation of the city by means of his city mental model (CMM).

1.7 Urban planning as a co-operative process

Urban planning can be considered as a process implying several actors, and the role of information systems is to solve the urban problems or at least some of them. But who are these urban actors, what kinds of relationship do they have, how can an information system support their decision? What could be the structure of such a computer system?

1.7.1 List of urban actors

It is possible to draw a partial list of persons having some impacts in the city evolution, varying from a single actor, for instance the city council, to all people living, working and having some links with a city. Facing those two extremes, a more effective list can include *de jure* and *de facto* actors (see Table 1.2).

1.7.2 Relationships between actors

Regarding the type of problems to solve, the relationships between two actors A_1 and A_2 can be:

- **in co-operation or synergy**; in this case both actors are working together to solve the same sets of problems, they agree on the same solution, they share the same action plans, and they pool their resources and means
- **in conflict**; in this case, the two actors have divergent interests in solving the problem; the conflict can be a conflict of objectives or a conflict of means, the more frequent being the conflict of objectives
- **in negotiation**; in this case, the actors know that their best interest is to work together within a limited common objective; they partially or temporarily agree to share some resources to solve a problem.

Generally speaking, among a set of actors, one or some are more important. Let us call 'dominant actor' the person who has the greater powers. In general, the dominant actor is the mayor, or the city council.

Table 1.2 Typology of urban actors

Names	Definition	Examples
De jure actors	• Actors appointed by an upper level of administration	• Local representatives of ministries
	• Actors elected by the people of the city	• City council, mayor
De facto actors	• Economic agents	• President of the Chamber of Commerce, CEOs of large local plants
	• Socio-political actors	• Presidents of some local associations, trade unions, political parties
	• *De jure* actors of neighbouring cities	
	• Some mob godfathers (*horresco referens*)	

35

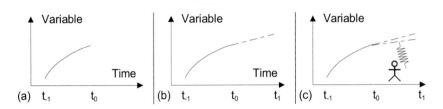

Figure 1.33. Evolution of an urban state variable (V) curbed by one actor. (a) Past evolution. (b) Forecast evolution when doing nothing. (c) An actor as a pulling force curbing evolution.

But in some cities, the owner of a very important plant can have a lot of power regarding the evolution of the city. For instance, in Seattle, Washington, surely Bill Gates has more power than the local mayor. Facing a mess of problems, some actors can co-operate for one single problem but are in conflict for another problem. So the previous typology must be used carefully because those relationships can vary according to the context and evolve over time.

Let us examine an urban state variable (for instance the number of crimes, or the number of unemployed persons, etc.). Figure 1.33a shows the past evolution, whereas Figure 1.33b depicts the evolution when doing nothing. Finally, Figure 1.33c illustrates the case when an actor is trying to pull the evolution of this variable according to his objective, based on some action plan. But in reality, there is not a single actor as a pulling force to curb the evolution of a state variable.

Figure 1.34a shows the example of two actors in co-operation trying to pull together a variable in the same direction, whereas Figure 1.34b gives the case of two conflicting actors. As a consequence the final evolution of a state variable will be the result of all actions of all actors (Figure 1.34c).

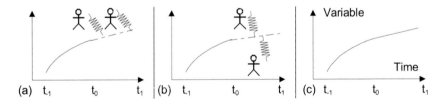

Figure 1.34 Evolution curbed by several actors and results. (a) Co-operative actors (in synergy). (b) Actors in conflict. (c) Evolution as a result of action plans carried out by several actors.

36

Figure 1.35 Layers of support for co-operative work.
According to Klein 1997.

1.7.3 *Information system as a tool for co-operative urban planning*

As a consequence emphasis will given on co-operation throughout this book, because it seems to be the best way to solve urban problems. However, we will always consider that some actors can be partially or temporarily in conflict. Figure 1.35 gives the main distinctions of some co-operative contexts. The lowest box shows communication between the actors and the top box shows their co-ordination.

According to Klein (1997), synergism looks more important than co-operation. Figure 1.36 shows a diagram stressing the key elements of synergism (conflict management, rationale capture and process management), together with their relationships.

Figure 1.36 Synergism.
According to Klein, 1997.

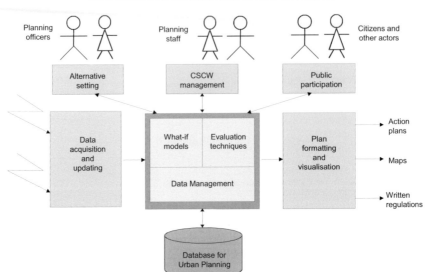

Figure 1.37 Sketch of an architecture for a computer system for urban planning.

1.7.4 Structure of a computer system for urban planning

Now we can give a sketch of a computer system dedicated to urban planning as depicted in Figure 1.37. If we compare it with a GIS structure given in Figure 1.27 and the DSS in Figure 1.10, we can see both similarities and differences:

- the acquisition model is similar to GISs and DSSs
- at output, not only cartographic visualisation is present, but also some graphic tools for exhibiting action plans, and editors for the generation of written regulations
- the kernel is similar to the DSS one, integrating some what-if models to simulate the development of the city together with some evaluation techniques, for instance, to estimate the costs and the efficiencies of some planning alternatives
- the great difference is the presence of additional modules:
  - **alternative setting**, which can also be present in a DSS; this module is activated by planning officials
  - **groupware management** or more exactly, computer supported co-operative work (CSCW) in order to organise and schedule the daily work of the planning staff, (see Chapter 8)
  - **management of public participation** in order to assist citizens in comparing alternatives, giving their opinions, synthesising them and so on (see Chapter 9).

1.8   Urban planning in different countries

The aim of this book is use by urban planners, practitioners or students, anywhere in the world. However, urban planning activities are organised differently in different countries. This book will assume that there exists, at municipal or metropolitan level, some service or agency in charge of city-wide planning in liaison with elected officials and politicians. More generally, by urban planners in the narrow meaning, we mean technicians (staff) in charge of designing planning activities, working for a municipality or a sort of city council. The wide meaning of planners will encompass also elected officials or officers appointed by an upper governmental level, perhaps regional or national, imposing some sort of master plan on local authorities. Examples will reflect mainly my experiences and knowledge in designing or assisting the design of information systems in some European and North American cities.

1.9   Conclusions

To conclude this introductory chapter, let me say that:

- information is the key element in any urban planning process
- the theory of the general system offers a nice conceptual framework to understand the role of information systems and also offers a design methodology which will feature in the next chapter
- sustainability is a key concept in urban planning; in other words, long-term planning is more important than short-term planning whereas the lifecycle of a city council is short term
- an information system for urban planning is not only based on a GIS, but needs also some other information to be integrated; in a sense, it is an essential part of the global information system of a local authority

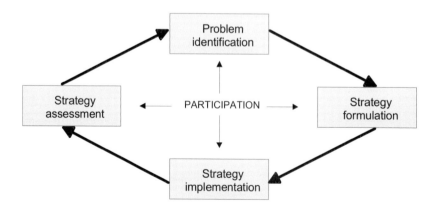

Figure 1.38 Brodhag wheel for strategy making and participation.
Source: Brodhag 1999.

- in order to reduce and solve conflicts, the idea is to build co-operative information systems by means of groupware technology emphasising collaboration between urban planners, local politicians and resident association spokesmen (Chapter 9)
- actions to solve or to reduce urban problems must be organised to form comprehensive action plans, and be consistent with the other urban actors (synergy).

Another way to conclude this chapter is to present the Brodhag wheel (Brodhag 1999) for strategy implementation and participation (Figure 1.38). Initially this wheel was made within the context of sustainable development but I think that for urban planning based on citizen participation, we have the same kind of diagram.

Having described the context of information systems for urban planning (ISUP), let us now describe some methodological tools to design an information system.

2

DESIGN METHODOLOGIES FOR INFORMATION SYSTEMS

For constructing an information system, several steps must be followed. Starting with the specifications, the first key step is to model the organisation and its context. For that we need a design methodology and the scope of this chapter will be to offer some methods for such a task. This chapter will give all elements for designing conceptual models as a key stage in urban information system building. The relational and object-oriented database models will be presented, together with methodologies to design them. Several small examples taken from different applications in urban planning will conclude this chapter.

2.1 Data, information, knowledge and metadata

Designing information system implies good definitions of the contents. The first step seems to properly define the vocabulary. For a start, the questions are what are the differences between data, information, knowledge, objects and metadata? Here only brief definitions will be given whereas Chapter 6 will detail knowledge and metadata.

2.1.1 Data

By data we usually mean a succession of digits or letters, i.e. a number, for instance 2.25, or a string such as 'Winston Churchill', without any semantic connotation. In other words, we do not know whether 2.25 represents a length, a temperature or a price. Similarly 'Winston Churchill' can be the name of a person, a cultural centre or a road. More generally, data are numbers, strings, Booleans, etc., that is to say elements for which one does not know the meaning, the origin, the validity over time. Some authors include also drawings, sounds, images and video into the data set. In these cases, we speak of multimedia data.

2.1.2   Information

Information is data that has been organised so that it has a meaning to the recipient. He can interpret the meaning and draw conclusions for making decisions and plan actions. Bearing this definition in mind, if one says 'length = 2.25 feet', one has got a piece of information, because in addition to a number, one has the meaning. By generalising, we can say that as soon as we have the definition of a data item, we have information.

2.1.3   Knowledge

By knowledge, we mean derived information about information. By analysing and synthesising information, we generate knowledge. So knowledge consists of data items that are organised and processed to convey understanding, experience, accumulated learning and expertise. Knowledge can be seen as the application of data and information to make a decision.

2.1.4   Objects

Since in computing, the word 'object' has a special meaning, we will try to avoid it when ambiguities can occur. Instead of speaking of real-world objects, we deal with features (sometimes named geo-features) existing in the world real or projected. Sometimes the word 'entity' will be also used. In computing, object means a set of bits/bytes which presents some unity. Generally, one computer object represents a single real-world feature, but in some cases object can also represent computer software items such as windows, etc.

2.1.5   Rules and metarules

Rules are often defined as knowledge about objects. So metarules are defined as rules about rules.

2.1.6   Metadata and metainformation

By metadata or metainformation, one means the definition of data names. When you say that 'p is the total population of London in 1850', this is a part of metadata corresponding to the definition of the value of p. Some authors often use the expression of data semantics by stressing the importance of keeping the signification of the data. More details will be given in Chapter 6.

2.1.7 From data to information, from running to managing a city

A functional definition of concepts can be attempted by saying that (Figure 2.1) as the clerks are using data daily in order to run the city, perhaps trying

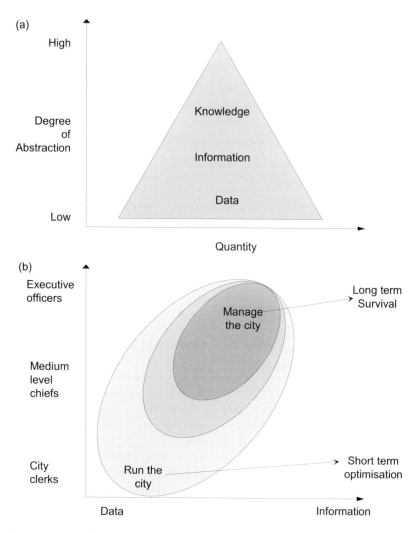

Figure 2.1 From data to information and knowledge. (a) Various levels of abstractions (from Turban and Aronson, 1998 with modifications). (b) From running to managing a city.
From Devlin, 1997 with modifications.

to optimise locally and at short term, the managers of the city need more global and synthetic information in order to ensure the long-term survival of the city.

In a different context (business), the expressions 'front office' and 'back office' are commonly used. As far as city management is concerned, front office can mean the clerks in charge of data collection and applying rules, whereas back office mean the definition of data, the creation or the discovery of rules and the management of metadata (top managers).

2.2   Modelling levels (external, conceptual, logical)[1]

As previously said, one of the main tasks is to model the organisation and its context. For many years, and especially since 1975, four different levels of data modelling have been used as a framework in the design of an information system (ANSI/X3/SPARC 1978). Starting from the so-called real world these are the:

- external level
- conceptual level
- logical level
- internal level.

Using Figure 2.2 as a global overview, let us look at each in turn. Here we first describe the characteristics of each level, before going on to elaborate only the external and conceptual levels. We see the levels as stages in the building of the database which is necessary and sufficient to accommodate the requirements of different people or organisations, or different uses by just one individual.

As a practical matter, the real world corresponds to a subset of reality that is of interest for the actors. Sometimes the notion of 'universe of discourse' is used. In this spirit it is assumed that one can build a database of a subset of reality, only if one knows how to describe it with words. However, in the multimedia world of urban information systems, much data, such as pictures, sketches or sounds, cannot be described by words only, but they can be described by other forms of digital or analog encoding. So we prefer to think of the universe of modelled phenomena. The very beginning of the database design process is the external modelling step in which potential users define their own subset of the real word, that is, what is relevant for their needs. Commonly, we deal with as many external models as we have different users.

The conceptual level corresponds to a synthesis of all external models. It is called this for two reasons: first, because it is made of very sound concepts; second, because it is the basis for the conception process. Although an abstraction of the real world, the result of conceptual modelling is supposedly quite concrete in nature, consisting of schematic representations of phenomena and how they are related. The organisation scheme created at this stage generally deals with only the information content of the database,

---

1   Most elements of this chapter come from Laurini and Thompson, *Fundamentals of Spatial Information Systems*, Academic Press, 1992, Chapter 9.

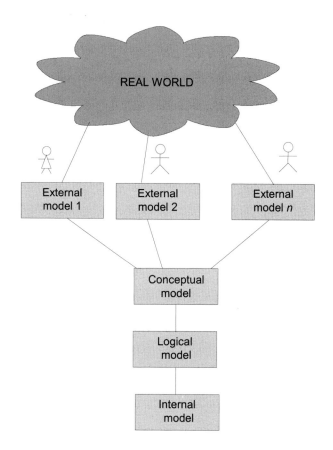

Figure 2.2 Data modelling levels (ANSI/X3/SPARC 1978).

not the physical storage, so that the same conceptual model may be appropriate for diverse physical implementations. The conceptual model not only provides a basis for schematising but is also a tool for discussion and, as such, a good conceptual model must be easily understandable by all potential end users, and more generally by all persons involved in the design process. In addition, in case of checking or evolution, it bears interesting characteristics for auditing (Batini *et al*. 1992).

This model sharing may be done by narrative statements, but the transfer to the next, logical stage, is easier if more formal mechanisms are used. Designers usually use the visually oriented entity-relationship approach in which phenomena and their associations are presented graphically as shown later. The representation of the conceptual organisation substantively has varied forms. Ordinarily it can be thought of as a sort of semantic data model in which the principal types of phenomenon and their associations and con-

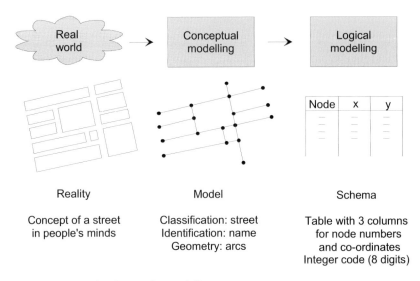

	Reality	Model	Schema

Concept of a street in people's minds	Classification: street Identification: name Geometry: arcs	Table with 3 columns for node numbers and co-ordinates Integer code (8 digits)

Figure 2.3 Example of steps for modelling streets. Bartelme 1996, with modification.

straints are laid out. External modelling uses similar tools but often consists simply of a narration of requirements stipulated by an individual in an interview process.

Internal models are at computer level, that is to say they represent the way by which information items are stored at bit/byte level. It is outside the scope of this book to give more details concerning this level (Elmasri and Navathe, 1989). In Figure 2.3 an example is given of steps for modelling some streets into a set of tables passing through a graph representation.

Many difficulties arise in designing the conceptual model as a synthesis of several external models, especially in dealing with the geometric data. Indeed, as illustrated later, external spatial models can be based on different geometric representations, and the synthesis can imply either of the following:

• having many geometric representations and then creating procedures to transform from one representation to another
• keeping only one model in the database and then transforming each external model, perhaps producing some difficulties for the computer programmers.

The next step in the design of an information system is the **logical modelling** phase. In fact, it is the first step in computing. The expression logical model has two meanings, and is not meant to convey that conceptual or external modelling are any less systematic. First, it constitutes a mathematical basis or a set of mathematical concepts. Second, it corresponds to the transformation (mapping) of the conceptual model with the tools offered by the logical modelling. In other words, we have a transfer between the conceptual models and a new modelling level which is more computing oriented. At the moment the logical models are either the **relational model** (Section 2.4) the basic structure of which is a table, or the **object-oriented model** (Section 2.7).

The realisation of a conceptual model is achieved by mapping the conditions of the data model into the definitions, constraints and procedures for one of these models, most especially today, the relational, or other newer types, extensions of the relational. In this way the permanent properties of the database are clearly specified, whatever the circumstances pertaining to a particular set of instances of phenomena. This procedure is associated, in a practical sense, in the creation of a data dictionary, a set of statements about important properties of the items, such as name, type of data, range of values or missing value.

The internal model is concerned with the byte-level data structure of the database. It explains every pointer and any information to access and handle correctly all data items. Whereas the logical level is concerned with tables and data records, the internal level deals with storage devices, file structures, access methods, and locations of data. For usual commercial database management systems, this level is not generally accessible to users.

2.3   Conceptual modelling: the entity-relationship approach

Conceptual modelling in database design often makes use of a formal approach known as entity-relationship modelling. First presented comprehensively in 1976 (Chen P. 1976), but based on some older ideas, it is a means to organise and schematise information. We first present the formalism and then give an example. In this formalism the basic components are:

- entities
- classes of entities
- relationships between entities or classes of entities
- attributes for both entities and relationships
- cardinalities of relationships
- integrity constraints.

An entity can be a person, place, thing or event. Mr Brown's parcel, the building 404, the car numbered 40HP69, the flight LH4093, are entities in the real world. But entities can be regrouped into classes. Relationships, definable for single entities or classes, are the associations between phenomena, such as the land parcel has a house, the car has an owner, and the flight goes between Buenos Aires and Tokyo. Both entities and relationships can hold attributes,

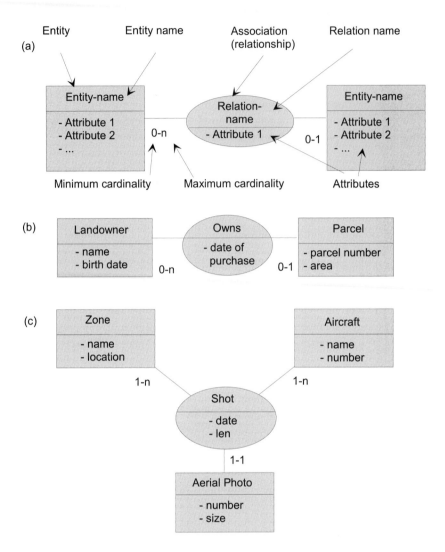

Figure 2.4 Examples of entity-relationship diagrams. (a) Nomenclature for entity-
relationship diagrams. (b) Example of a binary relationship. (c) Example of
a ternary relationship.
From Laurini and Thompson (1992).

their special characteristics, such as size of parcel, make of car, or flight
duration.

Generally, an association is binary (Figure 2.4a), that is linking only two
entities, but it may be more involved (Figure 2.4c). The degree of relationship
expresses the number of linkages. Cardinalities are expressed by four numbers

defining the minimum and maximum number of entities occurring in a relationship, in both the forward and reverse senses. Usually the minima are 0 or 1; when the maximum is unknown, we again use the letter n. In some special cases, we may need to use a summary statistic, for example, the average number of parcels per person. An integrity constraint can be defined as a predicate (value or symbol) which must be matched in order to confer integrity onto the model. We can apply integrity constraints to attribute values or to attribute definitions, but the most important are cardinality constraints. For details, refer to Laurini and Thompson (1992).

Let us take a simple example dealing with landowners and parcels. A landowner can own several parcels (minimum 0 and maximum n, which is often unknown), and parcels can be owned by no one or one or more (joint owners) people. So, the entity-relationship diagram can be designed as in Figure 2.4b. As attributes in this example, we can mention the landowner's name, his or her date of birth, and so on, the parcel number, its area. We can also consider attributes in the relationship such as purchasing dates.

In the visual convention we are using (Figure 2.4a), one of several in customary use, an entity or class of entities is depicted by a rectangle and a relationship by an ellipse (some authors prefer diamonds) linking two or more entities or entity classes. Attributes are shown as lists in the rectangle boxes, or attached to the relationship ellipses. Figure 2.4c illustrates the case of a ternary relationship among three entity classes of zone, aircraft and aerial photo. More generally, it is occasionally necessary to deal with n-ary relationships, where n denotes the number of classes.

A more comprehensive example, Figure 2.5, relates to a city containing several city-blocks, streets and parcels. The city can have several blocks and streets. A block consists of one parcel at the minimum and can be surrounded at the minimum by three streets. Conversely, a parcel only belongs to a unique city block (Figure 2.6). In a typical conceptual model, it is usual to deal with scores of entities and associations. Thus, the conceptual level's statement of the complete description of a future database is very useful for discussions among people in order to check the completeness and correctness, and to bring forth modifications. During the conceptual level step it is advisable to create a data dictionary, one component of metadata (see Chapter 6), incorporating the definitions of entities, of relationships, and of their attributes. This further facilitates the discussion between all potential users, allowing a designer to detect entities having contradictory definitions or different entities having the same definition.

2.4   Logical modelling: relational databases (Date 1987)

Now let us turn our attention to the logical modelling of a database using the relational model first. In this section we give only a small introduction to a well-known topic first formally examined in 1970 (Codd 1970). A relation is

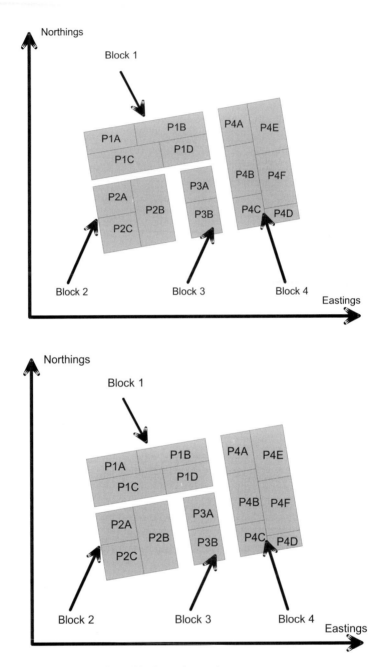

Figure 2.5 An example of city blocks and parcels.

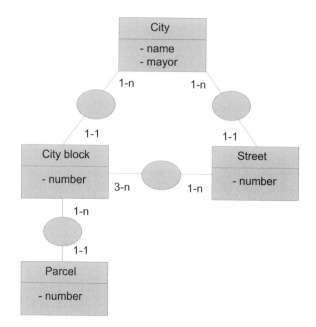

Figure 2.6 The entity-relationship diagram for land parcels and city blocks: a city can have several city-blocks and streets. Each block is surrounded by a minimum of three streets and can have a minimum of one parcel.

an organised assembly of data that meets certain conditions. While all relations are tables, not all tables are, strictly speaking, relations when the formal definition of this structure is recognised. A relational database is a collection of relations represented by statements as to contents, and tables containing instances of the relations. The database has no *a priori* pattern of connections among the relations. The basic tool is a statement corresponding to a relation. Using a conventional form of expression, a single relation for land-use planning can be:

R *(Zone-number, Area, Landuse)*

where R stands for relation, and the parentheses contain the names of three data items or attributes. A relation has a collection of attributes or data items representing some property of an entity about which data are to be stored. A table containing an instance of this example relation is shown below (Table 2.1).

The table consists of rows, columns and cell entries. The columns, sometimes referred to as fields, contain attributes. The rows, sometimes called records or tuples, contain particular instances of an entity. The cells contain one or more values for an attribute for an entity. There does not have to be

Table 2.1 Example of a table for land use planning

R	Zone-number	Area	Landuse
	34	4578	commercial
	35	3471	residential
	36	9065	industrial

any particular order to either rows or columns. Using more technical terminology to distinguish between features of tables and characteristics of relations, the main components of the relational model are attributes, domains, keys, relations and tuples (Figure 2.7). For each attribute, a data type and a format must be provided, and when necessary a unit of measurement must be specified. For example, for space co-ordinates, the metre or the foot can be selected.

One or more of the attributes is specified as a **key**, requiring that there be at least one data item column containing unique values for each row. In practice, the key is an identifier like a social security number, geocode, country name, or an arbitrarily assigned identifier like polygon number. At times the key may be a concatenation of several separate identifiers, like a household number-person number-travel trip-number combination. There may be several identifiers or label attributes, not all of which may be used as keys. A complete descriptive statement, or schema, for a relation looks like:

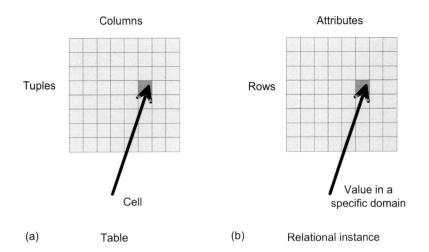

Figure 2.7 Nomenclature for relational tables. (a) Structure of a table. (b) Relational instance with a particular value.

Table 2.2 Example of a more complex relation

PURCHASE	Plot_ID	Owner_ID	Purchasing_date
	5678	45	13-03-76
	4567	67	16-05-81
	1208	85	01-07-78
	3089	60	20-12-80
	3218	49	31-01-82
	....	....	....

POINT *(Point-ID, X, Y, Accuracy)*

In this example, *POINT* is the name of the relation; *Point-ID, X, Y,* and *Accuracy* are attributes, each drawn from a specific domain:
  Another example of a relation schema is:

PURCHASE *(Plot_ID, Owner_ID, Purchasing_date)*

in which *PURCHASE* is the relation name, *Plot_ID* is the number of a plot of land (an attribute defined in an integer domain), *Owner_ID* is the identifier of the landowner (an attribute defined in a domain of character string), *Purchasing_date* is the date in which the previous landowner bought the plot of land (attribute defined in a date domain). Keys are often underlined. The row entries in the instance of a relation, the tuples, are shown in Table 2.2.

2.5 Spatial pictograms

In order to represent the spatial types of geographic entities in the entity-relationship approach, Caron *et al.* (1993) have proposed the use of four basic spatial pictograms which are Point, Line, Area and Volume as exemplified in Figure 2.8a. So any entity, in addition to classical attributes, possesses a spatial pictogram (Figure 2.8b) depending on its spatial type. Those basic pictograms can be combined. Figure 2.8c illustrates the cases of a complex entity (Water Supply Network) needing two spatial representations (pictograms) whereas the entity 'Working Site' depicts the case where two alternative spatial representations can be used, namely a point and an area. More precisely the semantics are the following:

- for simple pictograms, any occurrence of the entity 'Parking Lot' is represented by an area
- for complex pictograms, any occurrence of the entity 'Water Supply Network' is represented both at the minimum by a line or by a point; in some cases, it can be represented by using both pictograms
- for alternative pictograms, any occurrence of the entity 'Working Site' is represented more often by an area and otherwise by a point.

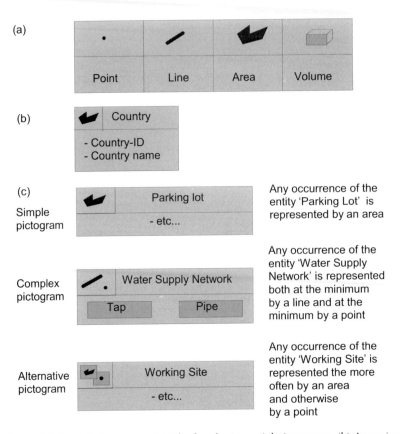

Figure 2.8 Spatial pictograms. (a) The four basic spatial pictograms. (b) An entity with its spatial pictogram. (c) Examples of simple, complex and alternative spatial pictograms.
From Caron *et al.* (1993) with modifications.

Figure 2.9 gives an example of an entity relationship diagram using pictograms (Caron *et al.* 1993).

Let me mention that Caron *et al.* (1993) also provide pictograms to deal with temporal aspects of entities and attributes (temporal pictograms).

2.6 Entity relationship model with entity pictograms (Lbath 1997)

In order to increase readability and abstraction, it is possible to replace the detail of an entity by a pictogram as exemplified in Figure 2.10. This idea will be re-used in the method OMEGA given in section 2.9. For instance special pictograms are used respectively for owner, road, town, plot and building entities.

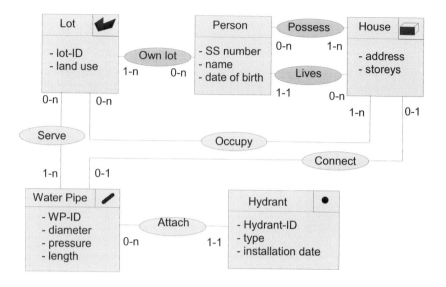

Figure 2.9 Example of an entity-relationship diagram with pictograms. From Caron *et al.* (1993).

2.7 Object-oriented spatial databases (Laurini and Thompson 1992)

The object-oriented approach is an attempt to improve modelling of the real world. Whereas previous modelling approaches were more record oriented, essentially too close to the computers, this new paradigm is a framework for generating models closer to real-world features. The ideal would seem to be to provide an homeomorphy, that is a direct correspondence, between real-world entities and their computer representation. This session gives an overview of the object-oriented approach and presents some small examples in urban information systems (Bertino and Martino 1993, Elmasri and Navathe 1989).

2.7.1 Classes, subclasses and instances

In an object-oriented database context, classification is a mapping of objects or instances to a common type. Objects can be regrouped into classes. By classes we mean a collection of objects with the same behaviour, that is responsiveness to some computing operations, or properties. We distinguish between the real-world functions like travel for commuters, or erosion of landscapes, which could be directly modelled, and functions or behaviours defined only for computer encoded data, like drawing a polygon. Instances are particular occurrences of objects for a given class. Within classes, it is possible to define subclasses: an example is the splitting of the class Driving Road

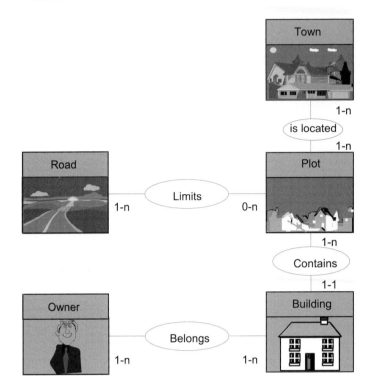

Figure 2.10 Example of an entity-relationship model with entity pictograms.

into two subclasses: Street and Motorway (Figure 2.11a). The resultant elements, using a conventional written formalism, are:

```
CLASS Motorway      : SUBCLASS_OF Driving Road
CLASS Street        : SUBCLASS_OF Driving Road
CLASS Turnpike      : SUBCLASS_OF Motorway
```

We also say that Motorway is a superclass of Turnpike, and Driving Road is a superclass of Motorway. In the sense class $\longrightarrow$ subclass we speak about specialisation of a class, and in the reverse sense, class $\longrightarrow$ superclass, we will use the term generalisation.

2.7.2 Attributes and data types

As an example, the attributes of a road can be its number, its width, its origin and its destination. For the turnpike, we have its toll. A more formal description including data types is:

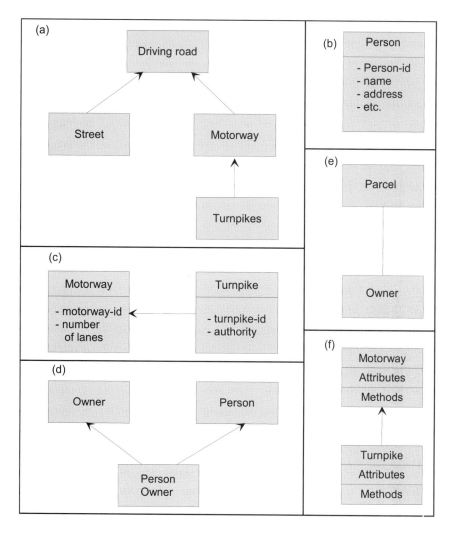

Figure 2.11 Example of object-oriented concepts. (a) Classes and subclasses. (b)
Attributes. (c) Attributes in inheritance hierarchy. (d) Multiple inheritance.
(e) Links between classes. (f) Inheritance of methods.

```
CLASS Turnpike      : SUBCLASS_OF Motorway
   {Turnpike-name   : string
    Toll-fee        : money}
```

A class is a sort of prototypical object defined with attributes. Instances of an
object have all those class attributes in common. Attribute values can be
defined at either the class or the instance level. For example, residents in a city

can have a value of `City-name` in common, but will have their own personal names. Sometimes, as a practical matter, a default mechanism can be used in a database in order to simplify assigning values for attributes.

Attributes are not only basic data types, in the sense of physical form, but also come from other classes. For example, we know:

```
CLASS City
  {City-name       : string
   Lord-Mayor      : person}
```

given that:

```
CLASS Person
  {Person-name     : string
   etc.
```

In some circumstances, combinations of attributes are necessary. In this situation, we can define a set or a list of attributes by `SET_OF` or `LIST_OF`:

```
CLASS City
  {City-name       : string
   Buildings       : LIST_OF Building}
```

Sometimes, we have a list of other instances of the same class:

```
CLASS City
  {City-name       : string
   Twin-cities     : LIST_OF City}
```

Such devices serve to link several objects while saving on data repetition. For example, lists of names can be referred to as needed from appropriate objects.

In object-oriented databases and in some extensions of the relational database model the concept of user-defined data types is encountered. Sometimes referred to as abstract data types (ADT), these will be defined for a particular need, possibly constrained by the limitations of the particular database system, reflecting the underlying programming language used. Generally, multimedia data are described by new abstract data types especially designed for the purpose of efficient defining and handling images, sounds and video information.

2.7.3 Inheritance

In classification hierarchies, an object in a subclass inherits all attributes of the corresponding higher-level class. In some situations, we need to work with classes as subclasses of several superclasses. That is, a strict hierarchical classification does not exist; instead there is a state of multiple inheritance (Figure 2.11c).

In some practical contexts, though, a `Person-owner` class is defined from the class `Owner` and from the class `Person`. In such a situation inheritance conflicts can occur. Indeed, supposing that both superclasses (Owner and Person) have attributes with the same name, but with different definitions or data types, there is a difficulty in choosing:

```
CLASS Person-owner : SUBCLASS_OF Person, SUBCLASS_OF
                     Owner
{Attributes        : .....}

CLASS Person
    {Person-Name    : string
    Date-of-bi      : date}

CLASS Owner
    {Owner-name     : integer
    Parcels         : LIST_OF Parcel
    Purchase-Date   : date}
```

In this example, a query for data for Date.Person_owner, leads to an ambiguity about the type since the two invoked possibilities, Date.Person and Date.Owner, have different data types.

2.7.4   Links between classes and instances

As at the semantic level, so objects may be linked in different ways in a database, and classes of objects may be linked, such as that between owner class and parcel class shown in Figure 2.11e. In this object level modelling, though, we do not customarily mention cardinalities; we begin with the conventional assumption that links correspond to general many-to-many relationships.

2.7.5   Methods

Objects are not only characterised by attributes but also by methods. We distinguish between the attributes (called declarative knowledge by some authors and which are the descriptive properties of the real phenomena as encoded) and the methods (the procedural knowledge, or some information as to what to do with those objects as encoded). The term method (sometimes called service) refers to an operation on the data, a procedure which can be applied to a class of objects. Thus we identify, and emphasise, the combination of data and operations that characterise the object-oriented model in contrast to the record-oriented model.

2.7.6   Examples

Figure 2.12 gives two examples of an object-oriented diagram. The first one (Figure 2.12a) describes very rapidly a city consisting of parcels and streets.

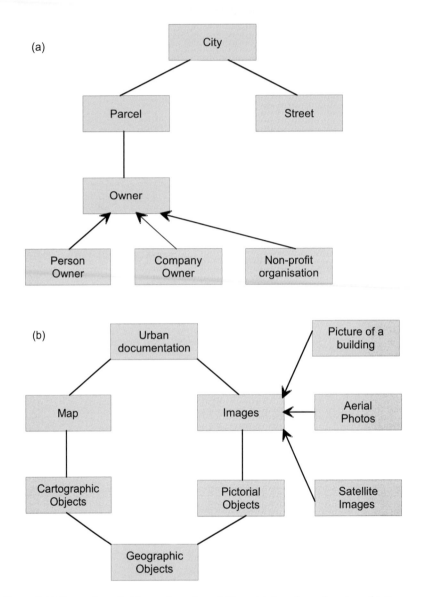

Figure 2.12 Examples of object-oriented modelling. (a) In urban planning. (b) For urban documentation.

Each parcel has one or several owners which can be either a person, or a company or a non-profit organisation (inheritance). A future database for urban documentation is given in Figure 2.12b in which the documentation consists of map and images. Maps regroup several cartographic objects which are linked to geographic objects. Similarly in images, pictorial objects can be found corresponding also to geographic objects. In an inheritance diagram, images can be aerial photos, satellite images and pictures of buildings.

2.8   Object-oriented database design methodology

Presently, there exist several methodologies in order to design an object-oriented database. Among those methodologies, apparently OMT (Object Modeling Technique) (see Rumbaugh *et al*. 1991, Derr 1995), and UML (Unified Modeling Technique) (Odell and Fowler 1998), are the most used. In comparison with the entity-relationship formalism, OMT/UML allows one:

- to define classes with attributes and methods
- to define inheritance
- to define aggregations and specialisation
- to define associations between classes
- and so on.

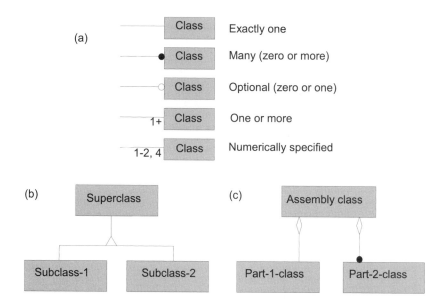

Figure 2.13 The OMT/UML formalism. (a) Various types of associations depending on cardinalities. (b) The specialisation in subclasses. (c) The aggregation formalism emphasising that Part-1-class has a unique instance whereas Part-2-class has several (Rumbaugh *et al*. 1991).

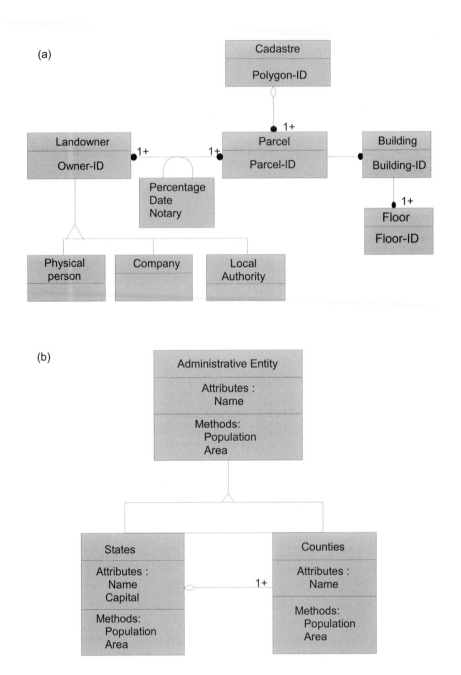

Figure 2.14 Example with OMT/UML formalism. (a) Ownership and cadastre without methods. (b) Administrative bodies in the US with their relations with methods.

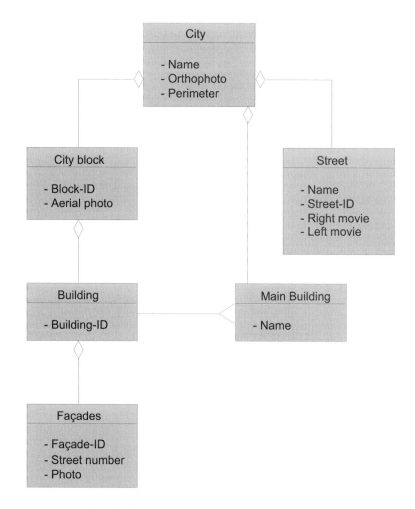

Figure 2.15 An urban database structure for storing façades of all buildings.

In Figure 2.13, the OMT/UML main concepts are given. Figure 2.13a gives the associations, 2.13b the specialisation in subclasses and 2.13c, the aggregation formalism. Figure 2.14a gives an example of using the OMT/UML methodology, one for cadastre (without methods) and Figure 2.14b illustrates administrative bodies in the US with methods. Remember that in the US, a city can belong to several counties. Figure 2.14a gives the territorial splitting of the US land. Here states are an aggregation, or precisely, technically speaking a tessellation (see Laurini and Thompson 1992), whereas in Figure 2.14b they appear within an inheritance diagram. In Figure 2.14a, the semi-circle (Percentage, Date, notary) represents attributes of a relation.

Figure 2.15 depicts an OMT/UML model for a database storing façades. More exactly, each building façade is stored as an image whereas two movies are stored per street, one for the left bank and one for the right bank (for the Cyclomedia system, see Figure 3.14). Such a database can be very interesting, for instance, to see the impacts of setting new electric or telephone poles and wires in a street.

In addition to information modelling, OMT/UML allows one also to model the behaviour of the information by means of a special notation for functional modelling and dynamic modelling. For those interested, please refer to Derr (1995).

2.9 Presentation of OMEGA (Lbath 1997, Lbath *et al.* 1997)

The OMEGA method, based on OOA (Coad and Yourdon 1991), MODUL-R (Caron *et al.* 1993), CONGOO (Pantazis 1994) and the unified method OMT-Booch (Rumbaugh *et al.* 1991) leads to an operational description of the graphic user interface, organisation and architecture. It covers all the aspects of design by considering four general views and giving an abstract level of a designer's point of view:

1 An **organisational view** includes requirements, collection of specific objects having semantic aspect for the end-user (use case) taking into

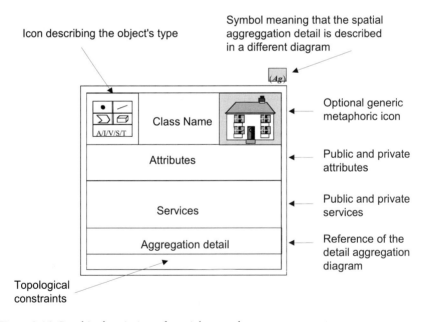

Figure 2.16 Graphic description of spatial meta-class.

account a possible existing alphanumerical system, organisation of costs, planning and staff structuring.

2   A **static view** describes objects and classes of objects with specific graphical representation (Figure 2.16); it is important to mention that for the pictogram representing geographic objects, one must consider the scale of the representation of the object (e.g. a road may be considered as a line or an area in function of scale's point of view). A meta-class is represented by a double frame with an iconic pictogram describing the object's type and a generic symbol giving visual information of thematic data (Figure 2.17). Each object class could have an end-user representation (icon) and technical representation (graphical representation with, in addition to the metaphoric icon, an iconic symbol for spatial data, the class name, attribute, and services).

  Each object meta-class has a multiple abstract representation depending on the designer requirements. Static view is associated to a class diagram model. A class diagram model is composed of a user class diagram in which objects are represented by icon metaphors (Figure 2.18) and a technical class diagram in which objects are represented by their technical representation (Figure 2.19). The user class diagram is more 'readable' for a non-expert end user; thus it will facilitate the validation of diagrams with end users.

3   A **dynamic view** includes the description of object reactions to different events. If the objects have a 'reasonable' number of transition states, each object state is associated with a dynamic metaphoric icon. This last point

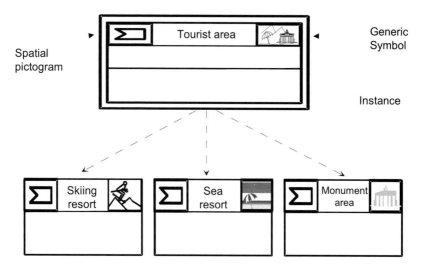

Figure 2.17 An example of a representation of meta-classes and their instances in OMEGA formalism (Lbath 1997, Lbath *et al.* 1997).

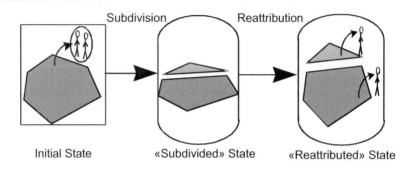

Subdivision   Reattribution

Initial State   «Subdivided» State   «Reattributed» State

Figure 2.18 Example of dynamic representation of object in function of its current state.

helps for testing (in different dynamic scenarios) and gives a semantic dimension for static scenarios (an example of a dynamic representation of an object according to its current state is given in Figure 2.18).

4   A functional view includes different services description and the main scenario description.

An example of a technical class diagram with the OMEGA formalism is given in Figure 2.20, in which metaphoric icons are replaced by a technical representation of objects.

2.10   Extended relational formalism

In some cases, it is interesting to extend the relation model in order to integrate some object-oriented characteristics. One of the more common is to introduce subtables in relational tables, i.e. instead of scalar attributes, there is an attributive table which is indicated by a star (*). So *R1 (a, b, c*)* means that the *c* attributes are a list of values; for instance for a person, we can write:

PERSON (Person_ID, Last_name, Christian_name)*

in which *Christian_name* indicates a list of several (perhaps one) first names. Similarly *OWNER (Owner_ID, Name, Address, Property*) Property* gives a list of properties. In some cases, not only one attribute constitutes a subtable, but several attributes. It is more complex when a subtable is nested in another subtable, such as: *R2(a, b, (c,d)*)* or *R3 (u, v, (w, (t,r)*)*)*. An example is modelling a person with his kids, who can have different toys, giving: *PERSON (Person-ID, Last_name, Christian_name*, (kid_date of_birth, kid-names*, toys*)*)*. Back to geoprocessing examples, we can easily write for a very simple polyline:

POLYLINE (Polyline_ID, (x, y))*.

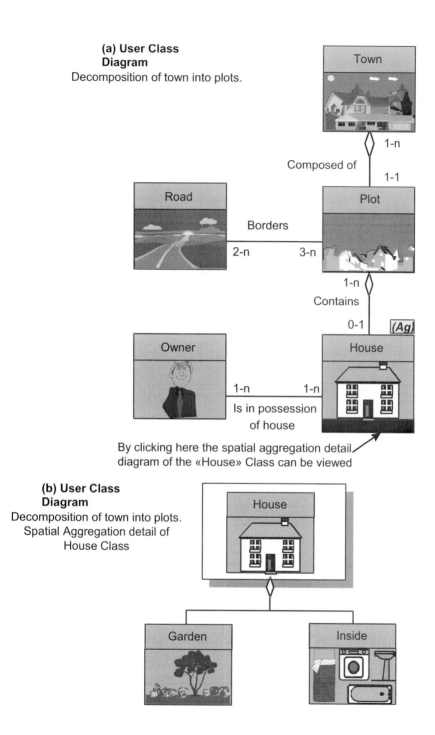

(a) User Class Diagram
Decomposition of town into plots.

Town

1-n

Composed of

1-1

Road

Plot

Borders

2-n   3-n

1-n

Contains

0-1   (Ag)

Owner

House

1-n   1-n

Is in possession of house

By clicking here the spatial aggregation detail diagram of the «House» Class can be viewed

(b) User Class Diagram
Decomposition of town into plots.
Spatial Aggregation detail of House Class

House

Garden   Inside

Figure 2.19 An example of a user class diagram with OMEGA formalism.

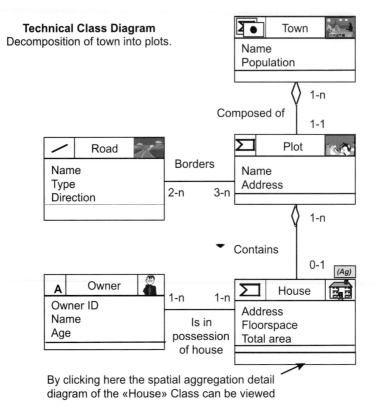

Technical Class Diagram
Decomposition of town into plots.

Figure 2.20 An example of a technical class diagram with OMEGA formalism. Metaphoric icons are replaced by a technical representation of objects.

2.11   Multirepresentation

Reality is very difficult to model, and several users can have different models of the same features. Consider the example of a street for which several representations can be offered (Figure 2.21):

• the officer in charge of transportation networks, or traffic control needs a model based on **graphs**

• for the officer of the cadastre, a street is defined by a **set of two lines** delineating the boundaries with the adjacent parcels

• for the officer in charge of pavement revetment, the street is defined by the **lines** along the curb

• the officer in charge of engineering networks needs a **3D model** to locate underground pipes

• etc.

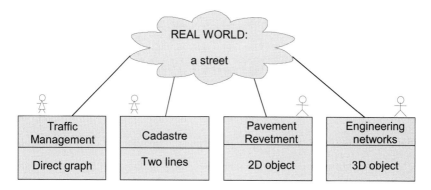

Figure 2.21 Multiple representations of a street: each user has his own model.

Moreover, the cadastre officer will perhaps have to deal with two representations, the legal one as given in the land titles, and the real ones as measures by the chartered surveyors.

Another very frequent case is a line with various generalisations: due to multi-scaling, the same line will be stored with different levels of details. At one scale corresponds a particular set of points. In any case, multiple representations will exist for several reasons, the low level being for the computer representation; indeed, each vendor can propose his own (seldom nicknamed proprietary) model of a polygon. Then at conceptual or semantic level, several representations can exist because of the diversity of applications which imply several ways of seeing the real world, with few or less details, with emphasis on some aspects and so on. The consequences of the multiple-representation will be drastic when exchanging data sets and in the context of interoperability of several GISs or computer systems.

2.12 Conclusions

The goal of this chapter was to give the reader the key methodological elements for designing an information system. Voluntarily, we have only stressed data aspects leaving procedural (methods) aspects less developed. For readers interested to know more about those aspects, please refer to Oestereich and Cestereich (1999).

3

DATA ACQUISITION IN AN
URBAN ENVIRONMENT

For designing an information system for urban planning, once given a preliminary list of data, there is a compromise in deciding between the cost and the quality of information, remembering that resolution is the more important key point. This third chapter presents a survey of the various techniques for data acquisition in order to develop any information system for urban planning.

A very important issue is the cost of data acquisition. For any geographical information system, generally speaking the cost of data amounts to five to ten times the price of hardware, software, orgware, staff training and maintenance. Even if at a very general level, no cost estimation had been carried out for any information system for urban planning, our feeling is that the ratio is more or less of the same order of magnitude. In other words, a city council needs to be very cautious in deciding whether or not to acquire some sort of data for its planning system.

As referred to in the previous chapters, different kinds of information must be integrated, not only geographical, but also from different sources, for instance from administrative forms, census, polls and so on. Of more and more importance are data coming from different types of vehicles and sensors, for instance for pollution measurement. Let us mention that we distinguish between alphanumeric and multimedia data. By alphanumeric, we mean numbers (integers or floats) and character strings; by multimedia, we refer to other kinds of data such as signals, images and audiovisuals. A sensor such as a thermometer sends alphanumeric data to a monitoring centre whereas a camera sends images or movies, i.e. multimedia data.

During the last decades, data acquisition has gradually passed from surveying to sensing. Indeed, surveying has been the more usual tool perhaps for centuries, but now, due to new technological developments, sensing and remote sensing techniques are gradually being employed in urban applications, especially the use of airborne or laser range systems.

This chapter will be organised as follows: first some descriptions will be given for administrative data. In the second part, conventional geographical data acquisition techniques will be described (digitising, theodolites and aerial

70

photos). Emphasis will be placed on novel vehicles for urban data acquisition. Then in the third part some techniques needing telecommunications such as the Global Positioning System (GPS) will be described, together with sensor-based acquisition techniques. We will conclude this chapter with some remarks concerning quality, scales and resolution in connection with applications.

3.1 Data from administrative routines

In urban planning, data coming from administrative sources are perhaps the more common. Among these, two different types can be examined, from censuses and from various administrative forms. In addition to those, data coming from polls will also been studied in this section.

3.1.1 Census data

For over a century in some countries, censuses have been organised in order to get data concerning the whole nation-wide territory. Generally speaking these data were defined for nation-wide needs but were captured at local level, and then aggregated to the national level and tabulated. In each country, several administrations have been created to collect those data. When the local level is compliant with the level of details required by urban planning, these data can be easily used also for local planning purposes. Once the disaggregation level is compliant, the next problem comes from the date of data collection of the census. Presently, it is common to organise censuses every ten years. Near the census date, the data can be reliable, but further on, sometimes some problems can occur due to their obsolescence and to the evolution of the territory for which the data was required. Generally speaking data regarding demography can be easily found in censuses. Facing the cost and the difficulties of organising censuses, some countries are thinking of organising rolling censuses.

3.1.2 Data from administrative files and forms

For some planning activities, administrative files can already exist. For instance for school management, a city can keep a file including information such as:

- name of the school
- address
- name of the head
- phone number
- number of classes
- number of pupils per class
- etc.

71

Such existing files can directly be used as input data to populate an information system for urban planning. More generally, all files storing data concerning public services can be used. For example, when somebody is applying for a building permit, he has not only a form to fill, but also to give some draft of the premises he wants to build. Such a form can be entered directly into the computer system and the draft can be scanned before being stored as an image in a database. More generally, many forms at municipal levels can be stored in a database.

3.1.3 Registers

In several countries, especially in Scandinavia, registers have existed for a very long time, sometimes for several centuries. By registers we mean files, or more exactly paper files which were kept and updated daily from their inception. Among them, the population register is the more commonly found. There are also registers for buildings, for enterprises, for land and so on. As an illustration, Finnish registers will be examined very rapidly.

According to Lahti (1994), the Finnish population census is based on state regulations first established for taxation purposes in 1556. Today a population register including address data is updated by public authorities constantly (for instance when a person moves from one place to another). To protect privacy, no information on an individual level can be divulged, except for special investigation purposes. Only aggregated, unidentifiable information and study results can be published.

For research purposes one of the most valuable elements of all databases and registers is their linkage data. The links enable the information in one source to be connected to information in other sources. The essential links in the Finnish database environment are co-ordinate data of buildings and personal identification codes (Figure 3.1). The databases are either nation-wide or local. The building data are updated constantly by the municipal building authorities along with building permits. They include information on building type, number of storeys, amount of floor space, main construction materials, heating systems, etc. All citizens are identified at some residential address and this is updated every time they move. The address data in turn can automatically be located in some physical building. These links allow some spatial analysis of a wide variety of demographic or other social or economic information.

It is important to mention that in many other European countries such as France, this kind of analysis is not possible because registers containing such information on individual citizens are not authorised by law except in special conditions. Practically, for planning purposes in France, the inception of such registers is forbidden. Similar registering systems exist also in Norway and Sweden. Where population registers exist, no censuses are needed.

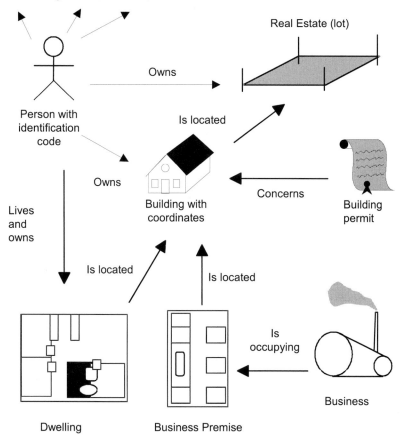

Figure 3.1 The basic elements of nation-wide databases and their internal and external links in Finland.
According to Lahti 1994.

3.1.4 Polls

In an ISUP, it can be interesting sometimes to store information from polls. Imagine a city having an important network of social service facilities, and that the mayor wants to get information about the number of people going or asking for services to those facilities, and their satisfaction level. For that purpose, a poll can be organised and the result can be integrated into the information system.

73

Based on these polls, some indicators of frequency and satisfaction can be designed in order to estimate or measure the efficiency of those services. Moreover those results can be used as a basis for the planning of the location of new social service facilities.

In a planning information system, poll data do not have the same status as other data, essentially because of their origin: they are date-stamped and will not be updated, and their accuracy bears some statistical characteristics; when a new poll is organised, the previous poll data are still kept or sometimes archived.

3.2   Map digitising and scanning

Existing maps are treasures but over time their quality is degrading. A very common way to begin acquisition is to start with existing maps. Presently, two acquisition methods exist, digitising and scanning.

3.2.1   Digitising

Digitising is a very common way to capture cartographic information. It is done by means of a digitising tablet (Figure 3.2) which is based on an orthogonal grid allowing the measurement of the co-ordinates in tablet pixels. After that, a scaling is done in order to obtain the exact co-ordinates. For this operation, it is necessary to put the map on the tablet so that the map sides are parallel to the tablet sides. If not, some future corrections will be necessary, especially in the form of rotations. Generally speaking, the procedure is as follows. First the operation captures the pixel co-ordinates and the real co-ordinates of four points, cautiously chosen in order to get the transformation

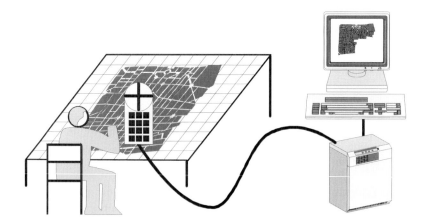

Figure 3.2 Principle of a digitising tablet.

parameters which will be applied to all co-ordinates. Let us call x, y the cartographic co-ordinates, generally in pixels, and X, Y the real-world co-ordinates. Since in urban planning, we generally do not take the earth curvature into account (for details, please refer to Laurini and Thompson 1992) we can write:

$$X = ax + by + c$$

$$Y = a'x + b'y + c'$$

In some cases, a more complex function must be used, such as the following:

$$X = axy + bx + cy + d$$

$$Y = a'xy + b'x + c'y + d$$

In more complex situations, some higher degree polynoms are used, or we refer to rubber-sheeting without or with constraints (see for instance http://www.gisparks.tas.gov.au/doco/genadoc7.2/map/gc_transform.html for a free rubber-sheeting product).

During this operation, known sometimes as Helmhert transformation, some problems can occur especially due to the difficulty of putting the mouse exactly at the same place as before. As illustrated in Figure 3.3, some additional corrections are necessary in order to confer the points' valid co-ordinates. Special methods exist to make the points A and A' exactly the same, and also so that the points D and G are not only the same, but also lie on the side KL. Similarly, overshoots and undershoots are problems very commonly encountered. This problem will be dealt with in Section 4.8.

The resulting format is called **vector format**. When the geometric data are not carefully organised into computer objects, in other words, when the spatial data can be seen as a set of points and lines without any connections between one another, the format is called spaghetti. Regarding spatial information computer format, please refer to Laurini and Thompson (1992).

3.2.2 Scanning

By means of a scanner, it is possible to capture spatial data. A map is put on the scanner, and all pixels are scanned. The output is a file of black and white pixels without any distinction. For instance a pixel can be part of a line, particles of dust, part of the text, part of the caption, etc. This is the well known raster format. In other words, an automatic capture is made without any intelligence. And the scope of any processing will be to recognise the constituent elements, such as segments, polygons, circles, texts and so on. And the next step will be to recognise perhaps streets, rivers, buildings, city limits and so on.

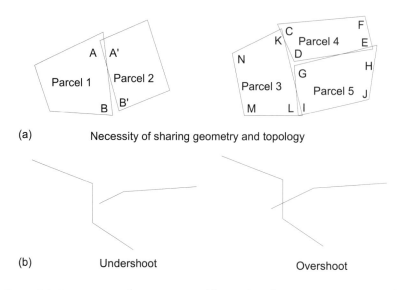

(a)   Necessity of sharing geometry and topology

(b)   Undershoot   Overshoot

Figure 3.3 Some commonly occurring problems when digitising. (a) Necessity of sharing geometry and topology. (b) Undershoot and overshoot.

In order to carry out this task, several intermediate steps must be taken.

• A step of vectorisation must be performed; generally speaking, a scanned line has a sort of width made of several pixels, and a line segment or a polygon is then represented by their kernel, often named skeleton; by vectorisation, we mean the procedure of finding the real core of this segment as depicted in Figure 3.4.

• The next is to start from those vectors to regroup relevant segments in order to represent a street, a building and so on; sometimes, one speaks about the skeleton of a line, or of a polygon.

It is necessary to explain that a set of special methods issued from artificial intelligence are used in order to recognise cartographic objects, namely pattern recognition. However, at the moment, all those tasks are not performed effectively, and because those methods are not efficient enough, more investigations must be carried in order to perfectly recognise all cartographic objects. The caption, taken as a set of pictorial rules can help a lot.

To conclude this way of capturing spatial data, let us say that optimism is included in this technique, but a lot of work must be done in order to get efficient and rapid procedures. Even so, map manipulation for scanning represents a short time, but recognition of cartographic objects will be very time-consuming, for several decades to come. It is likely that, scanning will be the main procedure for the future, followed by vectorisation.

Original segments

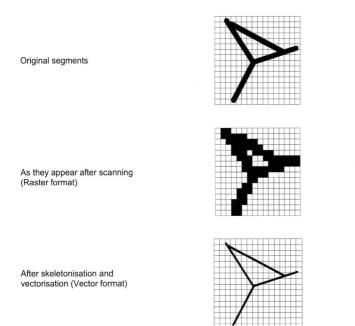

As they appear after scanning
(Raster format)

After skeletonisation and
vectorisation (Vector format)

Figure 3.4 From raster to vectors.

3.2.3   Terrestrial surveying

For centuries, terrestrial surveys have been made with chains. Now special devices such as theodolites and tacheometers are currently used in order to acquire terrestrial co-ordinates. These apparatuses measure only angles and distances, and starting from polar co-ordinates, Cartesian co-ordinates must be computed. Not only planimetry can be recorded, but also altimetry. In order to measure, some geodetic reference points must be used, perhaps a water tower, a church spire, etc. More generally, geodetic reference points are points which can be seen from a long way off, and all those reference points constitute a special network along which all measures are made. An example is given in Figure 3.5.

When one has to measure points which are badly located, i.e. when from the position of a theodolite, it is not possible to see both a geodetic reference point and the point to be measured, some additional reference points must be added. But the immediate consequence is to diminish the accuracy of the measurements.

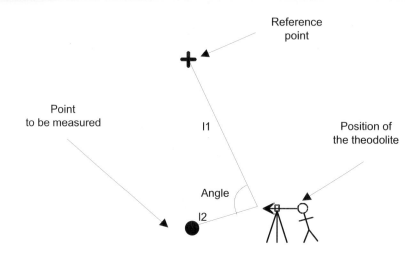

Figure 3.5 Principle of point measuring by means of a theodolite.

3.3   Aerial photographs and satellite images

Aerial photographs are also a typical way to capture spatial data. For this, an aircraft equipped with special cameras takes several photographs over a geographic zone (Figure 3.6). Photogrammetry is the name of the technique used to understand photos in order to extract information and especially measurements. We will finish this section by describing satellite images which bear some similarities to aerial photos.

The first aerial photos were made during the First World War. Originally, only black and white (grey) photos were possible. Now full colour photos are common. However, a controversy surrounds the choice between colour or black and white aerial photos. Some argue that the increased expense associated with colour film acquisition and laboratory development is not worth the resultant increase in interpretability. According to some aerial photo companies, the extra cost is generally less than one percent of overall project costs.

For urban applications, having a single set of distinct separated photos is not very interesting. People prefer to deal with seamless photos without distortions. The role of orthophotos is effectively to transform several photos into a sort of unique accurate map-like photo encompassing everything. We will finish this section by saying a few words about other kinds of photos and their use in geoprocessing. Eventually, we will examine the use of some vehicles which can be equipped with a camera to take photos along a street when being driven.

Figure 3.6 Example of an aerial photo.
Source: http://www.aerial-photos.com/vertical.htm. Photograph by CAP-Orlando, FL, USA. Published with permission.

3.3.1 Aerial photographs (Ciciarelli 1991, Falkner 1995 or Kraus 1993)

Aerial photographs are usually taken at a height of 2,000 feet to 10,000 feet, depending on the required level of detail (Figure 3.7). But height is not the only interesting parameter, scale is much more interesting. Typically scales vary from 1:50000 to 1:1000 and photo size is generally 23 cm × 23 cm. The following formula helps in defining the altitude H of the flight: $H = f \times m$, in which f stands for the focal length and m for the scale of the photos. For instance in urban areas, at 1:5000 scale, the soil surface covered by one photo is approximately 1.3 km². In order to prepare a flight, a line and a swath are defined in order to capture all spatial information usually with overlaps. Common overlaps are 60% longitudinally and 25% laterally (Figure 3.8).

The selection of the time of the year for organising a flight is very crucial. During winter, the trees have no leaves, and so the buildings, the footpaths, and all details are visible. But the shadows are more important. Moreover, vegetation analysis is quite impossible and often there is fog. In summer the contrary applies, shadows are limited, contrast is greater and vegetation hides some features partly or totally. Finally, in the northern hemisphere, in mean latitudes (Europe, US), the best period is from May to September during a fine day without clouds and near noon. Figure 3.9 gives an example of an aerial photo template for the Italian city of Bologna.

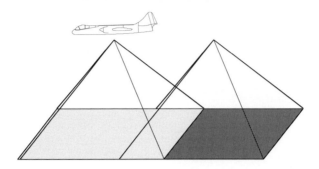

Figure 3.7 Aircraft taking photographs.

The selection of scale must be made bearing in mind the required level of detail. In urban zones scales are between 1:10000 to 1:20000, whereas in rural zones, scales range from 1:50000 to 1:20000.

3.3.2 Photogrammetry

The role of photogrammetry, or more especially the role of aerial photogrammetry is to extract accurate information from photos. Since the surface of the earth is not planar, the distance between geographic features and the camera is not regular. This phenomenon introduces distortions to the photos, and the role of photogrammetry is to provide methods in order to reduce, possibly to correct those distortions such as building leaning. Figure 3.10 shows an example of this effect in which the nadir corresponds to the vertical lines from the aircraft, or in other words, the centre of the aerial

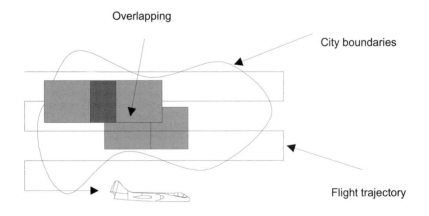

Figure 3.8 The flight trajectory, the swath and overlaps.

Figure 3.9 Example of aerial photo template for the City of Bologna.
Source: http://sit.comune.bologna.it/foto/foto97.htm. Published with permission.

photos. As another illustration of distortions, similar houses such as *A* and *B* in Figure 3.11 will appear to be different sizes.

Another difficulty in capturing data from aerial photos is the problem of what is the definition of a building; for instance, is a wooden hut a real building? From the sky you only see a roof and not what lies underneath. For centuries, surveyors used to take building co-ordinates on the surface, whereas in aerial photos, only roof co-ordinates can be measured (see Figure 3.12).

Another important problem is buildings appearing to be leaning; indeed, the aircraft is never at the exact vertical of the building, so buildings appear to be leaning in photos. In order to correct these distortions, some control points are defined on the ground by means of targets. A target is some kind of panel point that is placed on unobstructed ground prior to photography. They may take a variety of forms, resembling the letters 'T', 'Y', or 'X',

+ Nadir

Figure 3.10 Building leaning.

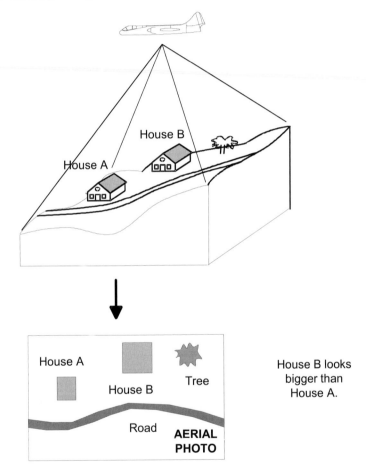

House B

House A

House A

House B

Tree

Road

AERIAL PHOTO

House B looks bigger than House A.

Figure 3.11 Similar geographical objects appear to be different sizes.

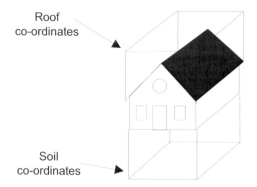

Roof
co-ordinates

Soil
co-ordinates

Figure 3.12 Differences between surface and roof co-ordinates.

painted white. Generally speaking, each photo requires a minimum of three control points.

Once photos have been made distortionless, the next step is the recognition of features. Usually, a person is in charge of analysing photos and of making the list of all existing land features with their characteristics, size, texture and so on. At the moment this task cannot be performed automatically by a computer. However, by image processing some elements can also be recognised. Regardless of the type of aerial photos, a wide range of features can be captured manually by an expert or perhaps semi-automatically. Sometimes this can be easily done after pre-marking these items with paint. This can be done for instance for manhole covers.

3.3.3 Orthophotos

As previously said, people prefer to work with seamless photos so that the collected photos look like a map with the same locational accuracy of a map with comparable scale; this is the role of orthophotos. By mosaicking different neighbouring photos, it is possible to construct orthophotos. For this purpose, generally cutlines are defined either manually or by performing a minimum distance algorithm. In order to carry this out, several steps must be performed depending on whether dealing with analogue or digital aerial photos. To perform an exact transformation in order to remove all distortions, a digital terrain model must be used (see Laurini and Thompson 1992). But in reality, sometimes this digital terrain model does not exist and some compromise must be reached, especially by using some premarked control points. As mentioned before, these points are generally white-painted marks evenly distributed. For all these points, co-ordinates and elevation are known at the surface and afterwards on the photos. A special analytic transformation is then determined and applied to all pixels.

The second problem is to remove overlaps. A first possibility is to cut to neighbouring photos midway. The great drawback of this method is to artificially cut features such as buildings. A solution is to cut photos not midway, but along a linear feature like a river or a road, especially along their axes. By doing this, the disadvantage of splitting surface features has disappeared but the main drawback is that in some places, the linear features used for splitting appear sometimes more or less wide than usual. By means of digital orthophotos some of these problems can be solved, starting from a rasterised or scanned aerial photo. A process called rectification removes the distortions arising from camera lens, the aircraft's position and elevation. **Pixel-by-pixel rectification** processes each pixel of the scanned image through photogrammetric equations. This rectification process can include:

- geo-referencing
- affine/polynomial transformations
- rubber-sheeting
- colour balancing
- etc.

One of the difficulties is storing orthophotos; sometimes some compression techniques are used, especially lossy compression techniques for which a compression ratio of 20:1 to 50:1 can be reached.

3.3.4 Satellite images (see Figure 3.13)

Satellite data are not very well used in urban contexts especially due to their resolution. Indeed present satellites like LANDSAT, SPOT, EURIMAGE or IKONOS deliver images the resolution of which can vary from 3 to 10 metres. In other words, this level of resolution does not comply with the requirements at urban level which are more or less one decimetre. However, for some applications, for instance green space planning in metropolitan areas, satellite data can help.

In the future, when satellite images will reach a resolution less than one foot, they will be interesting for urban planning purposes, especially as distortions will be very limited. Until this order of resolution is achieved, aerial photos will be preferred in the urban context. However, in land-use planning of developing countries, satellite images can help, especially because of their low price.

3.3.5 Photogrammetry and vehicles

Photogrammetrically speaking, spatial data are acquired not only by aircraft or helicopters, but also by other methods with cameras mounted in vehicles. Very briefly we will speak about façades which can be taken by different

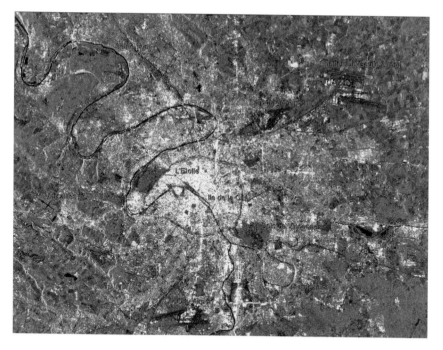

Figure 3.13 Example of a satellite image.
Published with permission. © ESA (1998) Original data distributed by Eurimage.

procedures such as Frank.

The Frank system developed by a Dutch company (Frank Data International) is based on a car-mounted camera taking fish-eye photos as shown in Figure 3.14. Starting from a set of fish-eye photos taken along a street, a very complex image processing routine can transform those photos into two films, one for the right-bank façade, and one for the left bank. For an example of the city of Rotterdam see Smit (1995).

3.4   Range finders and lasers

Systems based on laser-measures can be used for aerial surveys and also terrestrial surveys. In this section, only aerial surveys will be presented. The basic concepts of the airborne laser radar system are very simple (Figure 3.15a). A pulsed laser is optically coupled to a beam director which scans the laser pulses over a 'swath' of terrain, usually centred on the flight path of the aircraft in which the system is mounted. The round trip travel times of the laser pulses from the aircraft to the ground (or objects such as buildings, trees, power lines) are measured with a precise interval timer and the time intervals are converted into a range of measurements using the velocity of light. The

(a)

(b)

Figure 3.14 Example of a fish-eye photo and its transformation. (a) Cycloramas are digital panoramic images with a horizontal field of view of 360° and a vertical field of view from 30° below up to 60° above the horizon.
(b) CycloMedia is the brand name of the Frank Data products on the market.
Source: (http://www.frankdata.com/uk/frankdata.htm). Used with kind permission of Frank Data International N.V.

position of the aircraft at the time of each measurement is determined by GPS (see next section). Rotational positions of the beam director are combined with aircraft roll, pitch and heading values determined with an inertial navigation system, and the range measurements to obtain vectors from the aircraft to the ground points. When these vectors are added to the aircraft locations they yield accurate co-ordinates of points on the surface of the terrain (Figure 3.15b).

Typically, operating specifications permit flying speeds of 200 to 250 km/h, flying heights of 300 to 3,000 metres, scan angles up to 20 degrees, and pulse rates of 2,000 to 25,000 pulses per second. These parameters can be selected to yield a measurement point every few metres, with a footprint of 10 to 15 centimetres, providing enough information to create a digital terrain model

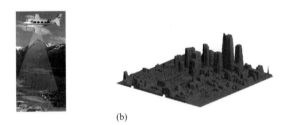

(a)   (b)

Figure 3.15 The principles of data acquisition with an airborne laser. (a) Example of an aircraft equipped with a range laser (Source: http://www.optech.on.ca/imagegall.html. (b) Example of terrain modelling in Denver, Colorado
Courtesy of http://www.eaglescan.com/products.html. Used with kind permission of Eaglescan.

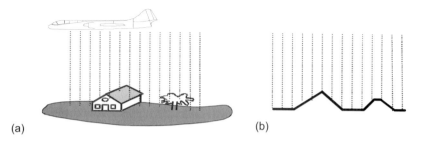

(a)   (b)

Figure 3.16 Using the airborne laser. (a) Regular beams are sent. (b) Result giving the profile.

adequate for many engineering and urban applications. Uncertainty of one-time measurement of geographical co-ordinates using this kind of device usually does not exceed 0.1% of flight height. The result is then the profile of the terrain as shown in Figure 3.16.

3.5   GPS (Leick 1995)

The American satellite system GPS (Global Positioning System) uses Hertzian waves sent to receivers (Figure 3.17b), enabling the system to compute their positioning on the earth with good accuracy. Knowing the signal celerity, it is possible to calculate the distance between the satellite and the receiver. By using several satellites forming a constellation (Figure 3.17a), it is possible to compute the exact position (longitude, latitude and altitude). The GPS satellites use a 20,200 km orbit, that is to say corresponding to a rotation period of 12 hours and have four atomic clocks. These clocks emit a signal at 10.23 Mhz. The life cycle of a satellite is seven years.

At present, 24 satellites have been launched and it will be possible to compute one's position at any location all around the world. But, when only two satellites are in view, the exact positioning cannot be defined and one has to await the arrival of a third satellite (Figure 3.18). The best case is when four satellites are in view, allowing drastic reduction of positioning uncertainties.

In absolute positioning, according to several correction modalities, the accuracy ranges from 20 to 100 metres; but in relative positioning (Figure 3.19), also called differential mode, i.e. when two receivers are used, the accuracy increases from 10 metres to 10 centimetres. Due to their precision, GPS receivers can also be used in geodesy in order to re-measure geodetic monuments (Figure 3.20).

Mobile GPS receivers can also be used to know vehicle positions. In those case, each vehicle's receiver is connected to a remote computer which stores information relative to its position and speed in real time. For several applications ranging from rescue vehicle, bus or ship transportation, GPS systems are currently in use. Two main kinds of path can be found for taking

(a)

(b)

GPS constellation

GPS receiver (Magellan Pioneer)

Figure 3.17 GPS constellation of satellites over the earth and example of a GPS receiver. (a) GPS constellation (Source: http://www.colorado.edu/ geography/gcraft/notes/gps/gif/orbits.gif). (b) Example of a Magellan GPS receiver.
Source: http://www.nvlt.com/pioneer.htm) Used with kind permission of Magellan.

co-ordinates in relative positioning: Figure 3.21a shows the case in which several points are measured and a point is used as a reference for differential mode, whereas Figure 3.21b, illustrates the trajectory of a vehicle taking several point co-ordinates. The first method is used to know the exact position of some geographic features, whereas the second can be used for instance to map a road. But in this latter case, the presence of tunnels can introduce some troubles.

3.6 Sensors

Due to the evolution, of urban planning, more and more data concerning urban environment are captured daily by means of sensors. For instance temperature, noise level, air pollution sensors are installed all over the city in order to take measurements in various locations generally in real time. These data are then sent to control centres in order to be regularly analysed.

88

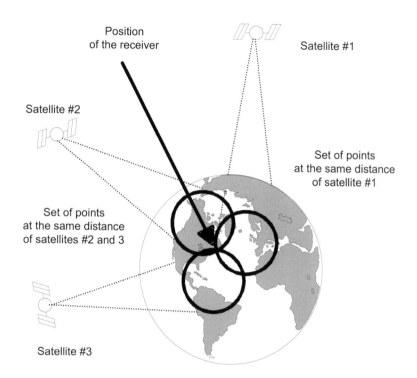

Position
of the receiver

Satellite #1

Satellite #2

Set of points
at the same distance
of satellite #1

Set of points
at the same distance
of satellites #2 and 3

Satellite #3

Figure 3.18 Computing location from three GPS satellites.

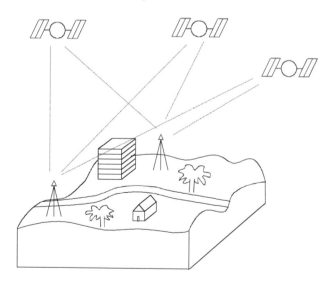

Figure 3.19 Relative positioning with differential mode with two receivers.

Figure 3.20 GPS field data collection.

Figure 3.22 depicts the architecture of a system for the monitoring of air pollution in a city. Various sensors are distributed all over the city (Figure 3.22a). These sensors measure some parameters and are linked to a special communication system in order to send data and receive orders. A possibility is to use cellular phones as a transmission device. Finally, Figure 3.22c gives the sketch of the complete system which is linked to a real-time mapping device giving animated cartography of air pollution. The sensors are initially set to send information every hour. If necessary, the control centre can call them back in order to modify the period of data sampling. These computer architectures will be studied in detail in Chapter 11. For some applications, sensors are mounted in vehicles. With this method, the vehicles' position is known thanks to GPS, and they are connected to a

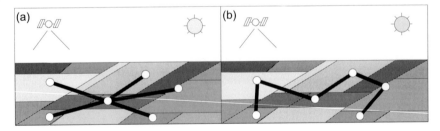

Figure 3.21 Different methods for capturing point co-ordinates with GPS. (a) From a central point. (b) Along a route.

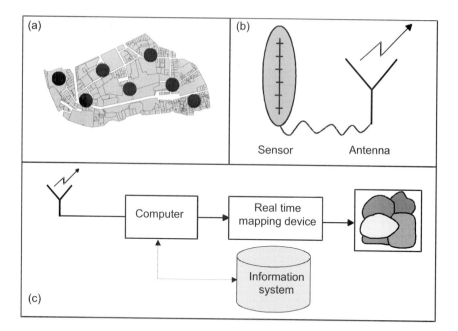

Figure 3.22 General architecture of an information system for air pollution control. (a) Location of sensors throughout the city. (b) Each sensor regularly sends measurements by means of a cellular phone to the control centre. (c) The control and monitoring centre receives phone calls from the sensors and produces a map of pollution which is regularly updated in real time.

control centre via another means of telecommunications as illustrated in Figure 3.23.

3.7 Voice technology and spatial data acquisition

Voice technology can help a lot in geographic data acquisition. Suppose somebody walking, or biking or driving is equipped with a voice device (Figure 3.24a) he can describe directly what he sees. By directly storing voice information, or even using voice recognition technology, spatial information can be stored easily, possibly in connection with GPS. The companies Datria (www.datria.com) and Stantec (www.stantec.com) are marketing this kind of device. Applications can be as follows:

- road maintenance and inventory
- wastewater and neighbourhood appearance
- building inspection
- abandoned cars and fines
- etc.

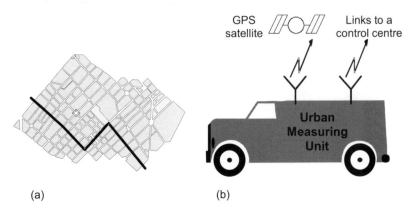

(a)   (b)

Figure 3.23 Principle of a vehicle equipped with sensors connected to GPS in order to make real-time measurements. (a) Path of the vehicle in real time. (b) Vehicle showing connection to GPS and to a control centre.

The main advantage of this technology is the huge reduction in the cost of data entry.

3.8 Remarks on quality, scales, resolutions and applications

The goal of this chapter was to give some hints regarding the acquisition of data for urban planning. As previously said, cost is the main problem in urban data acquisition. One of the main difficulties is selecting the mode of acquisition taking accuracy specifications into account. In reality, a sort of balance must be reached between the acquisition mode, the scale, the accuracy and the cost. Figure 3.25a gives some indications of the scale and the

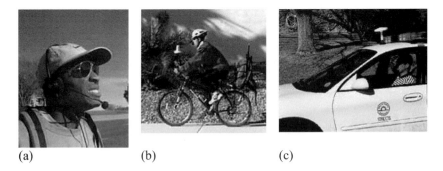

(a)   (b)   (c)

Figure 3.24 Example of voice-based technology in geographic data capture. (a) Headset microphone. (b) Biking unit. (c) Car unit.
Source: http://www.stantec.com/datria). Used with kind permission of Stantec Global Technologies Ltd.

acquisition mode, and Figure 3.25b the possible utilisation. Concerning urban planning, the scales vary from 1:1000 to 1:100000 and we can see that aerial photos are the medium which has more or less the same scale, that is to say, more relevant for urban planning applications.

To conclude this chapter, let us say that a contract must be signed between the user (city) and the data producer in order to agree some specifications especially regarding data quality. At the reception of data, the user must run some checking procedure in order to verify, at least statistically, that all specifications concerning data quality are met. Otherwise data must be

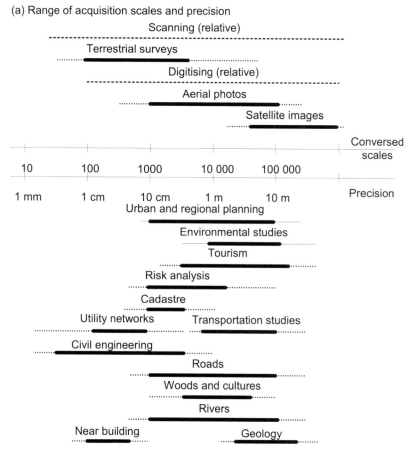

(a) Range of acquisition scales and precision

(b) Range of utilisation scales and precision

Figure 3.25 Scale, data acquisition and main utilisation. (a) Range of acquisition versus scale. (b) Range of utilisation versus scale.

rejected. This is one of the tasks of the GIS data custodian. In the following chapter, we will deal with the problems linked to quality control and updating of geographic information. However, due to the very strong level of geographic data consistency, some procedures can be run to detect topological errors and correct them.

4

QUALITY CONTROL AND MULTISOURCE UPDATING OF URBAN DATABASES

All geoprocessing applications need to cope with updated and good quality geographical databases in order to secure any urban planning decision. However, as updating the name of a new landowner is an easy operation in a land database, it is more difficult when one has to integrate spatial information, especially that coming from different multimedia sources. But before dealing in details with multisource updating, let us give some general information regarding quality control and quality assessment of spatial information.

4.1 Quality control and assessment of spatial information

Data quality can have different meanings. According to Veregin (1998), quality is commonly used to indicate the superiority of a manufactured good or to indicate a high degree of craftsmanship or artistry. We might define it as the degree of excellence in a product, service or performance. Moreover in manufacturing, quality is a desirable goal achieved through management and control of the production process (statistical quality control). Many of the same issues apply to the quality of urban databases, since a database is the result of an acquisition and production process, and the reliability of the process imparts value and utility to the database.

According to the Glossary of Quality Term,[1] one can distinguish:

- **Quality control**: the operational techniques and activities used to ensure that quality standards are met. Generally speaking these quality standards are defined in the specifications.
- **Quality assessment**: the operational techniques and activities used to evaluate the quality of processes, practices, programs, products and service.

Concerning geographic data, several studies were originally made by Guptill and Morrison 1995 and Antenucci *et al.* 1991, and so now conventional spatial data quality components are as follows:

---

1 Glossary of Quality Term: http://www.pinellas.k12.fl.us/qa/glossary.htm

- lineage
- accuracy
- resolution
- feature completeness
- timeliness
- consistency
- quality of metadata.

It is important to mention that with time data quality rapidly diminishes if no maintenance action is taken (Figure 4.1a); however, when regular maintenance actions are undertaken, the quality will always be near the top quality level (Figure 4.1b). In some very drastic cases, a sort of total database re-engineering must be carried out: to examine all data and enrich them. This case is very important when some novel applications are to be installed and require some higher levels of data quality. For more details regarding geographic data quality, please refer to the special chapters of Longley *et al.* (1999).

4.1.1 Lineage

According to Guptill and Morrison (1995) lineage is any information that describes the source observations or materials, data acquisition and compilation methods, analyses and derivations that the data sets have been subject to. In other words, lineage information defines the origin and the history of the concerned data sets. Of course, when necessary updating dates and methodologies must be mentioned. Generally speaking, lineage must include: (a) source data description (b) transformation information, and (c) input/output information.

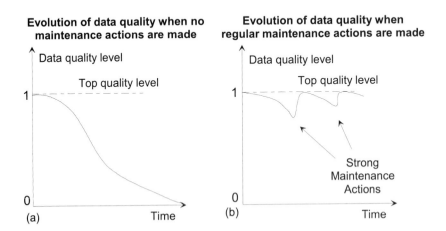

Figure 4.1 Quality and time. (a) Evolution when no maintenance is undertaken. (b) Evolution when regular maintenance actions are performed.

4.1.2 Accuracy

Accuracy is the inverse of error. Many people equate accuracy with quality but in fact accuracy is just one component of quality. An error is a discrepancy between the encoded and actual value of a particular attribute for a given entity. 'Actual value' implies the existence of an objective, observable reality. Accuracy is always a relative measure, since it is always measured relative to the specifications. To judge fitness-for-use, one must judge the data relative to the specifications, and also consider the limitations of the specifications themselves. We can define:

- **Spatial accuracy** as the accuracy of the spatial component of the database. The metrics used depend on the dimensionality of the entities under consideration. For points, accuracy is defined in terms of the distance between the encoded location and 'actual' location. Error can be defined in various dimensions: x, y, z. For points, often spatial accuracy can be presented in terms of a spatial tolerance such as ± 1 foot. For lines and areas, the situation is more complex. This is because error is a mixture of positional error (error in locating well-defined points along the line) and generalisation error (error in the points selected to represent the line). The ϵ band (Blakemore, 1983) is usually used to define a zone of uncertainty around the encoded line, within which 'actual' line exists with some probability (see Figure 4.2).
- **Temporal accuracy** as the agreement between the encoded and 'actual' temporal co-ordinates for an entity. Temporal co-ordinates are often implicit only in geographical data, e.g., a time stamp indicating that the entity was valid at some time. Often this is applied to the entire database (e.g., a map dated '1995'). More realistically, temporal co-ordinates are the temporal limits within which the entity is valid (e.g., Pothole 25CHAM-50 existed between 2/12/96 and 8/9/96).
- **Thematic accuracy** as the accuracy of the attribute values encoded in a database. The metrics used here depend on the measurement scale of the

Figure 4.2 Examples of error band for a point, a line and a polygon.

data. Quantitative data (e.g., precipitation) can be treated like a z-co-ordinate (elevation). Qualitative data (e.g., land use/land cover) is normally assessed using a cross-tabulation of encoded and 'actual' classes at a sample of locations.

4.1.3 Resolution

Resolution (or precision) refers to the amount of details that can be discerned in space, time or theme. Resolution is always finite because no measurement system is infinitely precise, and because spatial databases are intentionally generalised to reduce details. Resolution is the aspect of the database specifications that determines how useful a given database may be for a particular application. High resolution is not always better; low resolution may be desirable when one wishes to formulate general models.

Resolution is linked with accuracy, since the level of resolution affects the database specifications against which accuracy is assessed. Two databases with the same overall accuracy levels but different levels of resolution do not have the same quality; the database with the lower resolution has less demanding accuracy requirements. (For example, thematic accuracy will tend to be higher for general land use/land cover classes like 'urban' than for specific classes like 'residential'.) It is possible to define spatial resolution, temporal resolution and thematic resolution. See Veregin (1998) for details.

4.1.4 Feature completeness

Feature completeness is the degree to which all database features of one type are captured corresponding to the data specifications (Guptill and Morrison 1995). Completeness refers to a lack of errors of omission in a database. It is assessed relative to the database specifications, which define the desired degree of generalisation and abstraction (selective omission). There are two kinds of completeness (Brassel et al. 1995):

- **Data completeness** is a measurable error of omission observed between the database and the specifications. Even highly generalised databases can be 'data complete' if they contain all the objects described in the specifications.
- **Model completeness** refers to the agreement between the database specifications and the 'abstract universe' that is required for a particular database application. A database is 'model complete' if its specifications are appropriate for a given application.

As an extension of completeness, let us mention the avoidance of duplicates, because it appears that sometimes one object can be erroneously duplicated, or even have several copies.

4.1.5   Timeliness

Timeliness addresses the currency of feature representations and their respective attributes (Antenucci *et al.* 1991). In other words, timeliness is an aftermath of updating which will be described extensively in this chapter.

4.1.6   Consistency

According to Veregin (1998), consistency refers to the absence of apparent contradictions within a database. Consistency is a measure of the internal validity of a database, and is assessed using information that is contained within the database. **Spatial consistency** includes topological consistency, or conformance to topological rules, e.g., all one-dimensional objects must intersect at a zero-dimensional object. This aspect will be partly described later in Section 4.8. **Temporal consistency** is related to temporal topology, e.g., the constraint that only one event can occur at a given location at a given time. **Thematic consistency** refers to a lack of contradictions in redundant thematic attributes. For example, attribute values for population, area, and population density must agree for all entities.

4.1.7   Quality of metadata

The quality, consistency and completeness of metadata are also components of data quality. Let us speak about **metadata quality**. More details will be given in Chapter 6.

4.1.8   Veregin matrix

To measure quality as a whole, Veregin and Hartigai (1995) have proposed the matrix shown in Table 4.1 allowing the combination of Space, Time and Theme with Accuracy, Precision, Consistency and Completeness. For instance, Arnaud (2000) used this matrix for assessing the quality of postal addresses in Portugal.

4.1.9   Cost of quality maintenance

During the lifecycle of a database, it is necessary to maintain a high level of quality, especially by regular updating. Figure 4.3 depicts a balance which has on one side the cost of quality maintenance, and on the other the induced cost when no maintenance is done. The big problem is that the maintenance cost is generally admitted to be around 10% per year of the acquisition cost, which corresponds to a huge amount of money. But on the other hand, it is very difficult to estimate the cost of decisions made with erroneous data. However, I think that the maintenance cost is lower.

Table 4.1 Veregin and Hartigai (1995) data quality matrix

	Space	Time	Themes
Accuracy			
Precision			
Consistency			
Completeness			

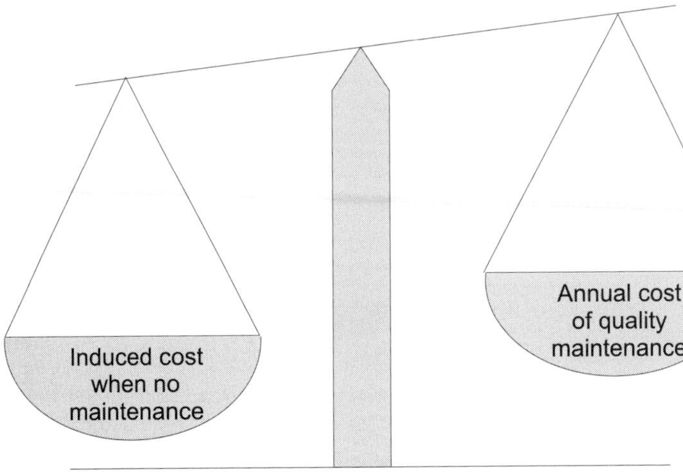

Figure 4.3 Balance between the annual cost of quality maintenance, and the induced cost when no maintenance is done.

In Roberts (1999), the reader will find some recommendations for building a plan for regular GIS data maintenance, and sharing maintenance costs with the different users. Moreover, he insists on the necessity of maintaining good quality metadata. For a comparisons of standards for quality, please refer to Aalders (1999).

4.2   Generalities about updating[2]

In geoprocessing, updating can have different scopes linked to the evolution of real objects or the evolution of the stored knowledge we have about them. More precisely, there exists:

---

2   For a preliminary version of this chapter please refer to Laurini (1994a).

- updating because new objects are created, or existing objects have evolved or disappeared
- updating for correcting measurements or topological errors
- updating in order to integrate more accurate co-ordinates, to confer more geometric details to simplified objects or to convey them more semantics
- updating in order to align several different layers of data
- updating in order to extend the coverage.

Moreover, integration can differ from the various sources of acquiring data, that is to say:

- newly made measures with more accurate devices (e.g. theodolites) with similar scopes
- vector and raster format, for instance, aerial photos or satellite images
- various data producers, using different bases or standards or different specifications
- etc.

In addition, sometimes also for the same sources, data formats can differ. However, in this chapter, this aspect will not be studied.

Geographical data have the particularity that topology is hidden underneath geometry; so, by reasoning about data which are already stored in a database, some other data can be derived by means of mathematical procedures. In this case, one can speak about an internal source for enriching data in comparison with external enriching. Finally, this updating can be performed by different methods such as alphanumeric, visual or automatic updating. The scope of this section will be to give some information regarding updating, especially by using visual interfaces for controlling multi-source integration within an urban database. Let us examine five cases (Figure 4.4):

- when the original is supposedly wrong and new corrected information must replace a subset of it; for instance, when a city-block was re-measured by means of newly purchased devices with better accuracy and when these measurements must be integrated without disturbing the remainder of the existing database (local updating by geometric refinement)
- when all the geographical information is supposed to be a little bit wrong and we want to correct the data everywhere by using, for instance, aerial photos, or to check whether the topology is good; we call this task global updating
- when the original database is supposed to be right and some new information must be added at the neighbourhood; a classic example is when adding a new city-ward into an existing spatial database (database coverage extension) for which some boundary matching must be performed
- when the database consists of several layers of data and that some alignment is necessary

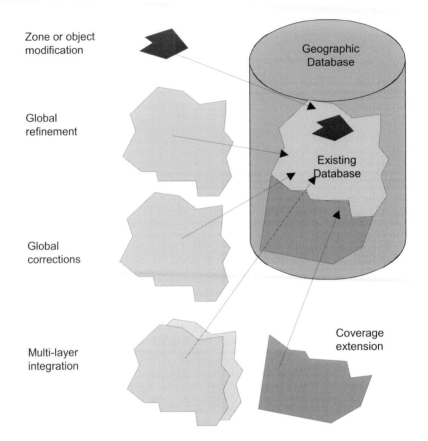

Figure 4.4 Different cases of geographical database updating.
Reprinted from *Computers, Environment and Urban Systems* 18 4, R. Laurini 'Multi-Source Updating and Fusion of Geographical Databases' 243–56 © 1994, with permission from Elsevier Science.

- when measurements or topological errors have been detected and corrected; in this case, let us call this kind of modification, global corrections; in this case, the source is internal.

In this chapter, we will not solve the general problem, but only give some hints using an example. In the following, we will suppose that the database has the following relations (extended relational formalism Section 2.10):

R1 (#parcel,(#segment)*)
R2 (#segment, (#point1, #point2)*)
R3 (#point, x, y)

102

4.3 Alphanumeric updating

By means of a language such as SQL,[3] it is very possible to update manually an alphanumeric database. For instance, suppose that we are obliged to change the co-ordinates of a point say $\#point = 2537$, $x = 4567$ and $y = 7890$, one can write:

```
UPDATE R3
SET x=4567,
    y=7890
    WHERE #point=2537;
```

Now, suppose that we want to confer more detail to a simple segment (657) in order to replace it by two new segments as depicted in Figure 4.5.

First, we have to know all the identifiers and the new values and in this case, we have to run the following transaction in SQL:[4]

```
BEGIN
DELETE FROM R2
    WHERE #segment=657;
INSERT INTO R3 (#point, x, y)
    VALUES (NEW.#point,760,6640);
INSERT INTO R2 (#segment, #point1, #point2)
    VALUES (NEW.#segment,120,CURR.#point);
INSERT INTO R2 (#segment, #point1, #point2)
    VALUES (NEW.#segment,CURR.#point, 121);
COMMIT;
END;
```

So, very rapidly, it appears that manual updates via alphanumeric methods are not a good way to follow, and in this case, visual interfaces seem to be a possibility.

4.4 Zonal updating and refinement

As previously explained, updating must be performed in two cases, when there is an evolution of geographic objects and when there is an evolution of the knowledge about them. In zonal updating, there are several cases to consider according to side-effects. Indeed, due to errors in measurements, sometimes some modifications must be carried out at the vicinity of the zone to be updated.

---

3 For details on the language SQL (Structured Query Language), please refer to http://www.jcc.com/SQLPages/jccs_sql.htm

4 In SQL, the prefix CURR and NEXT correspond respectively to current and newly-created attributes; COMMIT is necessary to conclude a transaction.

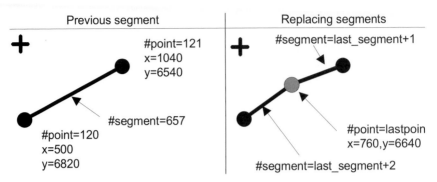

Figure 4.5 Replacing one segment by two segments.
Reprinted from *Computers, Environment and Urban Systems* 18 4, R. Laurini 'Multi-Source Updating and Fusion of Geographical Databases' 243–56 © 1994, with permission from Elsevier Science.

Another case of updating is when a subset (patch) of the database is considered as erroneous and when new measures are intended to replace old measures especially with more details. This problem is also known as geometric refinement (which appears as the opposite of generalisation). This case presents some elements in common with the previous case except that sometimes some compromises must be made between new measures and older information. Two main cases have to be studied, the integration of new measures and the integration of scanned maps, but those cases can present similarities. However, the more interesting dichotomy concerns the necessity of local corrections based on elastic transformations (rubber-sheeting)

4.4.1 Object integration without side-effects

The first case of multi-source updating is when receiving a new file to integrate. For instance suppose that a building permit file is arriving, giving new building co-ordinates, and that the scope is to integrate this new file into the cadastre file. The situation is as shown in Figure 4.6. In this case, updating is very straightforward.

4.4.2 Updating with local modification without elastic transformation

The other case of updating is when some topological modifications and perhaps geometric ones must be performed at the vicinity. For instance, let us rapidly examine the integration of a new road into a cadastre (Figure 4.7). In this case, the parcels located at the vicinity of the newly constructed road must be reorganised. This action can be done either automatically or by means of a visual interface. In this case, it is not necessary to move some points or some

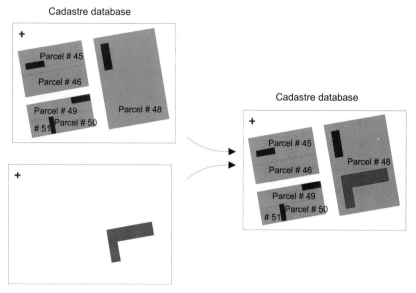

Figure 4.6 Integrating new building file into cadastre.
Reprinted from *Computers, Environment and Urban Systems* 18 4, R. Laurini 'Multi-Source Updating and Fusion of Geographical Databases' 243–56 © 1994, with permission from Elsevier Science.

segments in order to match some control points; elastic transformation is not necessary.

Figure 4.8 illustrates the case of cadaster updating (Spéry 1999). Figure 4.8a features the initial situation, in the parcel P_6 is split into two parcels P_{61} and P_{62} with new measures. When there are no modifications in the surroundings (French-style case), this approach will lead to 'patchy and dirty' databases, because some topological inconsistencies may exist (Figure 4.8b). But in the Danish-style cadaster (Figure 4.8c) the neighbouring parcels are corrected accordingly (Enemark 1998): there are topologically clean. For instance, P_3 is still a four-sided polygon within the French case without any modification, whereas within the Danish case, it becomes a five-sided polygon in which two previous points are modified. When using a GIS product including a topological spatial model, the Danish-style modifications are easy to implement.

4.4.3   Updating with elastic transformation

In some more frequent cases, the geometry of the patch to integrate does not meet the existing geometry implying the necessity to perform an elastic

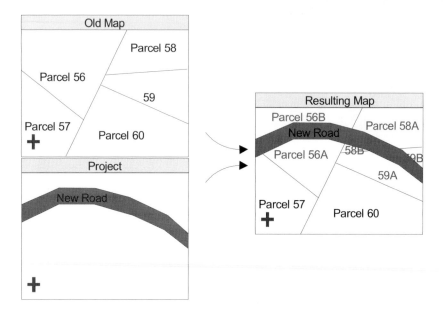

Figure 4.7 Principle of local updating with topologic reorganisation and without elastic correction.

Reprinted from *Computers, Environment and Urban Systems* **18** 4, R. Laurini 'Multi-Source Updating and Fusion of Geographical Databases' 243–56 © 1994, with permission from Elsevier Science.

transformation or rubber-sheeting at the vicinity. To perform these operations correctly, in addition to classical creation operation, elementary actions can be as follows:

- force-fit a point (co-ordinates)
- force-fit the length of a segment

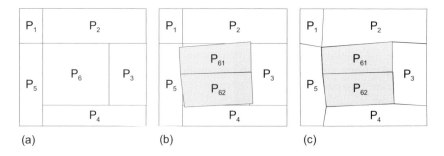

(a)   (b)   (c)

Figure 4.8 Consequences of introducing updating into cadastral maps (after Spéry 1999). (a) Initial situation. (b) Integrating without corrections, French-style approach. (c) Integrating with correction, Danish-style approach.

- force-fit an angle (especially right angles)
- force-replace a segment by a new polyline
- etc.

In other words, this transformation must follow some constraints (see Figure 4.9).

In the case of using scanned patches, a recognition phase must be performed. The problem is not only to detect segments and characters, but also to recognise cartographic objects such as houses, fields, streets, rivers and so on. For this scope, the caption can act as a graphic rule base. In some cases, the transformation must be performed not only on the zone itself but also at its neighbourhood (elastic zone). Under the common names of map conflation (Marx and Saalfeld 1988), rubber-sheeting, etc. several techniques can be used in order to perform feature alignment and feature matching (Laurini and Thompson, 1992).

As related in the previous chapter (Section 3.2.1), if we call x and y the co-ordinates before rubber-sheeting, and X, Y after, the simplest way is to use the linear formulae such as:

$$X = axy + bx + cy + d$$
$$Y = a'xy + b'x + c'y + d'$$

In more complex cases, some polynomial functions will be used. In addition, a very special aspect is the updating when multiple representations of a

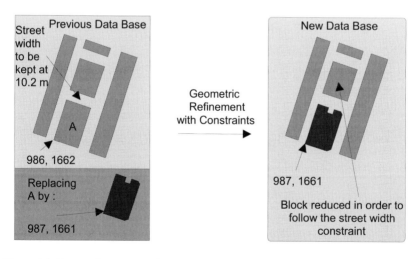

Figure 4.9 Zone refinement with constraints.

Reprinted from *Computers, Environment and Urban Systems* 18 4, R. Laurini 'Multi-Source Updating and Fusion of Geographical Databases' 243–56 © 1994, with permission from Elsevier Science.

single geographic object (for instance a street) are used. In this case, transferring updates from a representation to another is not very easy.

4.5 Global updating

When a reference system has to be changed or when new measures of your local geodetic points are provided, the whole database co-ordinates must be corrected accordingly. A similar case can appear after a very violent earthquake modifying the appearance and the location of all geographical features. For performing this kind of task, several methods exist:

- conventional rubber-sheeting when a few control points are provided (Figure 4.10)
- more sophisticated rubber-sheeting based on several points with constraints
- global updating based on aerial photos.

In this section, let us examine more precisely the case of using aerial photos for global updates (Photopolis Project, Servigne *et al.* 1991, Servigne 1993).

The idea (Figure 4.11) is to organise flights regularly (for instance every other year) and automatically to compare photos and database contents (Figure 4.12). In order to perform this global update, the following tasks have to be undertaken:

- *image dewarping*, in order to correct distortions due to relief, camera tilt, perspective effects together with removing of photo overlaps
- *construction of a texture and a knowledge base*, in order to store texture modification due to some changes in season, hour, vegetation, etc.

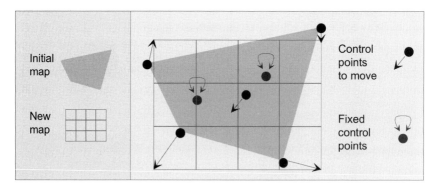

Figure 4.10 Multi-point rubber-sheeting with constraints.
Reprinted from *Computers, Environment and Urban Systems* 18 4, R. Laurini 'Multi-Source Updating and Fusion of Geographical Databases' 243–56 © 1994, with permission from Elsevier Science.

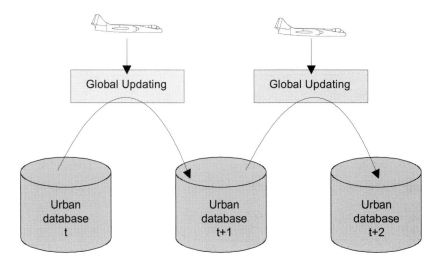

Figure 4.11 Main steps for regularly updating an urban database from aerial photos. Reprinted from *Computers, Environment and Urban Systems* 18 4, R. Laurini 'Multi-Source Updating and Fusion of Geographical Databases' 243–56 © 1994, with permission from Elsevier Science.

- *extraction of zones with homogeneous textures*, and comparison with the database contents
- whenever a discrepancy is detected (for instance, new building) an update will follow
- sometimes some extra information must be used such as building permit files.

The main results of the Photopolis project are the following:

- if the creation of a territorial database is not possible from aerial photos, its update is possible because it is model-based
- erroneous soil building co-ordinates update is possible when the error is greater than 0.50 m (especially due to the presence of roof overhangs)
- not-yet-stored buildings detected by this method can be put into the database but their legal status must be checked (clandestine buildings)
- necessity to deal with a well-done knowledge base in order to facilitate successive updating flights (Figure 4.11).

The procedure is as follows: an analysis of the photo is made together with the cadastre, based on a triangulation. Then the contours of objects are calculated by the snake method (Kaas *et al.* 1987) in order to give double contours which collapse into a single contour. The pictorial objects are detected and finally they can be compared with the cadastral objects. See Figure 4.13.

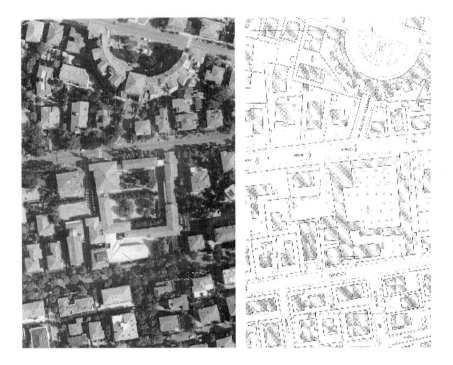

Figure 4.12 Example of aerial photos and the corresponding data in the database. The goal is to compare the map and the photos in order to find the differences and to update automatically the database (Tellez and Servigne 1998).

4.6   Mixing two layers

When one has to mix several layers, updating problems will occur. For instance suppose that there are two different databases, one for water supply and another for gas supply, both managed by different companies and measured by different surveyors (Figure 4.14). If no corrections are performed, some discrepancies will occur. Consequently, modifications must be performed with or without constraints. In case of constraints, the user must indicate them, for instance, by means of a visual interface.

In those cases, some elastic transformations with constraints must be performed implying a force-fit to some points, segments and angles. Generally speaking when extra information does not exist, a 50%–50% transformation must be used. One aspect which is not examined in this chapter is the appearance of sliver polygons (see for instance, Lester and Chrisman 1991).

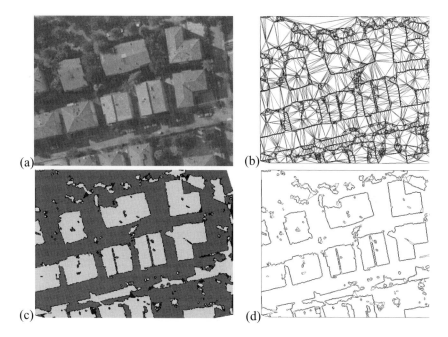

Figure 4.13 Object extraction in aerial photos. (a) Initial Photos. (b) Triangulation before colour classification and contour points extraction. (c) Triangles classification bright/zone, dark/background, black/boundary. (d) Objects extraction with double contours (Tellez and Servigne 1998).

4.7 Coverage extension

Generally speaking, when extending the coverage of a spatial database, due to the differences in measurements, we can have discrepancies at the boundary. For instance, at the boundary of the United States and Canada, the discrepancy between database objects can be more than 50 metres. So the problem is to make compromises in order to create a new single boundary (Figure 4.15). Some methods exist, but the main drawback is that they also distort all objects at the neighbourhood of the boundary. For instance, a rectangular house is slanted, a rectilinear road becomes a piece-wise road, etc.

The task for avoiding this kind of drawback is to implement a transformation process. This is an elastic transformation similar to rubber-sheeting, but with constraints. These coefficients must be determined by the least-squares method based on some control points with constraints. Possible rules are the following:

• If the boundary segments of A are considered more accurate than those of B, then keep them (A) and force-fit the boundary segments of B;

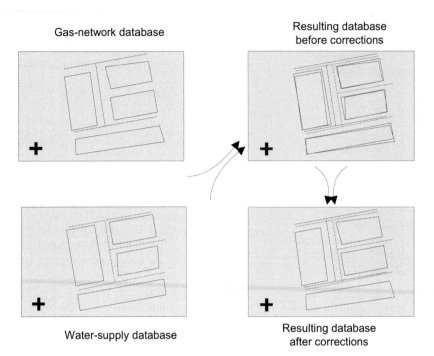

Gas-network database

Resulting database
before corrections

Water-supply database

Resulting database
after corrections

Figure 4.14 Example of layer mixing.
Reprinted from *Computers, Environment and Urban Systems* 18 4, R. Laurini 'Multi-Source Updating and Fusion of Geographical Databases' 243–56 © 1994, with permission from Elsevier Science.

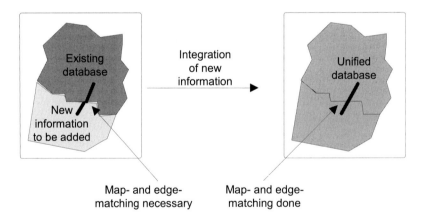

Figure 4.15 Geographic database coverage extension.

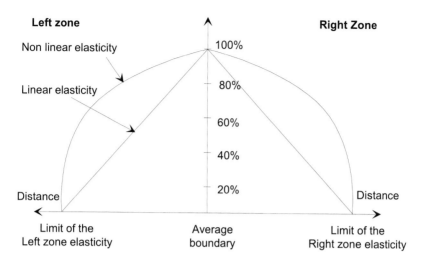

Figure 4.16 Elastic functions.

• If boundaries are different and inaccurate then take a sort of mid-line and distort objects neighbouring the boundary accordingly.

Similarly, the user has to select the way of performing the elastic transformation at the boundary with or without constraints. Here the constraints are:

• alignment of streets
• parallelism of curbs or parcel limits
• rectangularity of some buildings.

A second very important issue is the range of this transformation. The transformation must be important at the vicinity of the boundary, but vanishing with the distance (Figure 4.16). So a swath (Figure 4.17) is defined each side of the boundary. In order not to deal with continuity problems, it was decided to let the user define the limits of this swath in the middle of streets. Indeed, by this method, transformations of visual effects are drastically reduced.

4.8 Global corrections

Due to the excessive price of digitising, many geographical databases are not well captured and structured: even if the maps look good, in reality, the database is likely to consist of a set of lines of various lengths with no sets of cartographic objects (called spaghetti databases). Some experts claim that 80% of actual geographical databases are sets of disordered text fragments and spaghetti lines instead of well-constructed geographical objects. Apparently two situations exist, single-layer and multiple-layer corrections. Let examine each of them.

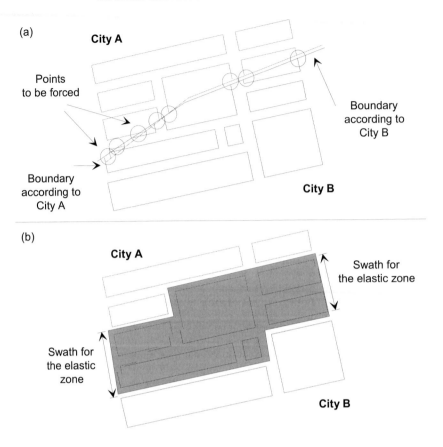

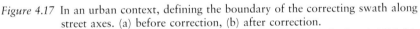

Figure 4.17 In an urban context, defining the boundary of the correcting swath along street axes. (a) before correction, (b) after correction.

Reprinted from *Computers, Environment and Urban Systems* 18 4, R. Laurini 'Multi-Source Updating and Fusion of Geographical Databases' 243–56 © 1994, with permission from Elsevier Science.

4.8.1 Single-layer corrections

As an introductory example, let us examine a water supply network: the lines look good, the pipes and taps are well located. Let us add some houses which are connected to the water supply system. Let us now assume that for some reason, some pipes must be turned off for repair. By topological analysis, a well-constructed system will immediately show the houses without water. But unfortunately, due to topological inconsistencies, often this kind of operation cannot be performed.

In the existing databases, there exist several inconsistencies which are shown in Figure 4.17. Let us explain them. In some cases, it can be seen that the lines do not snap together, because of undershoots or overshoots. Some lines are shown twice and some nodes do not match. This is because the acquisition of those data was done manually. So, it is now apparent to the reader that a topological reorganisation is a basic necessity before launching into any spatial analysis or spatial reasoning.

Figure 4.18 depicts an example of topological errors in the network, and Figure 4.19 in the case of a cadastre. Figure 4.20 shows some errors due to bad finishing during digitising (clicking many times instead of once). The case of topological correction is often a case of updating since the origin is not a different source (additional information from outside acquisition), but an internal one (that is to say only existing information stored in the database). Indeed, by extensively using geometric and topologic properties, some kind of consistency checking can be performed implying the correction of the database. For more details see Laurini and Milleret-Raffort (1991).

Figure 4.21 emphasises the case of contour curves for which some constraints exist so that we can base some correcting reasonings on them. Generally speaking, a step h is chosen (for instance 10 metres) and the set of all

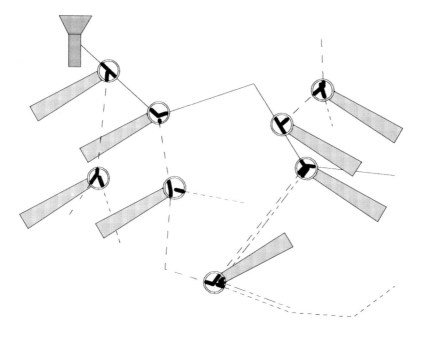

Figure 4.18 Detecting topological errors in a water network.

Reprinted from *Computers, Environment and Urban Systems* 18 4, R. Laurini 'Multi-Source Updating and Fusion of Geographical Databases' 243–56 © 1994, with permission from Elsevier Science.

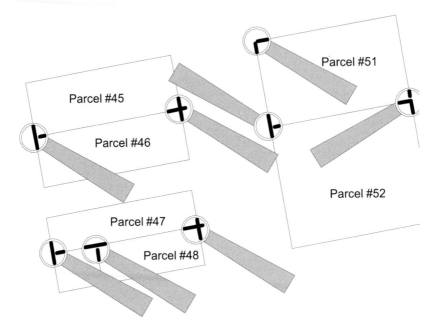

Parcel #51

Parcel #45

Parcel #46

Parcel #52

Parcel #47

Parcel #48

Figure 4.19 Detecting topological errors in a cadastre.
Reprinted from *Computers, Environment and Urban Systems* **18** 4, R. Laurini 'Multi-Source Updating and Fusion of Geographical Databases' 243–56 © 1994, with permission from Elsevier Science.

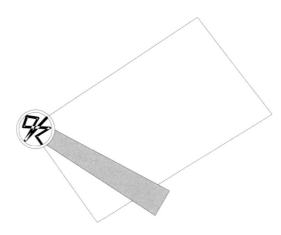

Figure 4.20 Strange errors due to awkward finishing of digitising (Ubeda *et al.* 1997).

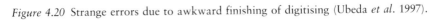

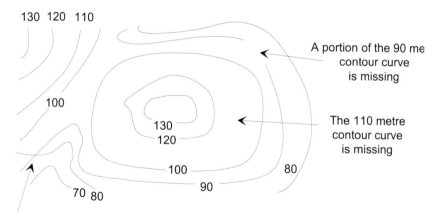

The 80 metre curve intersects the 70 metre curve

Figure 4.21 Detecting topological errors in contour level curves.
Reprinted from *Computers, Environment and Urban Systems* 18 4, R. Laurini 'Multi-Source Updating and Fusion of Geographical Databases' 243–56 © 1994, with permission from Elsevier Science.

contour curves has a tree structure. With the extended relational model (Section 2.10), we can write:

CONTOUR *(#curve, closing_flag, z, (x,y)*)*

Obviously, some inconsistencies can occur. Let us note *min_z* and *max_z*, the limit elevations. It is possible to consider the following integrity constraints:

- *Existential constraints:* root and leaves apart, if z does exist then $z + h$ and $z - h$ must exist.
- *Including constraints:* root and leaves apart, any contour curve z must include the curve $z + h$ and must be included in the curve $z - h$
- *Non-overlapping constraints:* any two level curves must not intersect.
- *Open curves* except at the boundaries, level curves are never open.

So, by using those constraints, corrections can be done as a special case of updating.

4.8.2 *Multi-layer corrections*

Another possibility for updating is by confronting several layers. For instance, suppose we have cadastre, building and water-supply layers, some inconsistencies can be discovered as shown in Figures 4.22 and 4.23.

At this step, two types of memberships can be distinguished:

- *nominal or entity membership*, that is to say that there is a link or a pointer between two entities

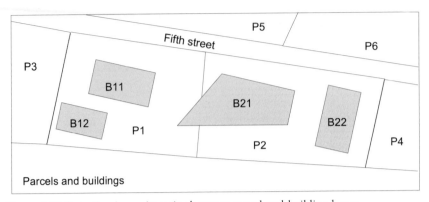

Figure 4.22 Detecting inconsistencies between parcel and building layers.
Reprinted from *Computers, Environment and Urban Systems* **18** 4, R. Laurini 'Multi-Source Updating and Fusion of Geographical Databases' 243–56 © 1994, with permission from Elsevier Science.

- *cartographic membership*, that is to say that the membership can only be seen on a map

4.8.3 Using spatial integrity constraints (Ubeda 1997, Ubeda and Egenhofer 1997)

Ubeda has recently proposed a methodology for checking the topological consistency of geographic databases based on spatial integrity constraints. They are the following based on the previous works of Laurini and Milleret-Raffort (1991) and Plümer (1996) concerning spatial integrity constraints:

- for each node, there is only one point

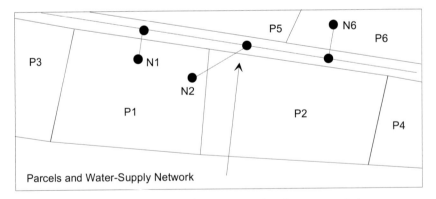

Figure 4.23 Detecting inconsistencies between parcel and water-supply layers.
Reprinted from *Computers, Environment and Urban Systems* **18** 4, R. Laurini 'Multi-Source Updating and Fusion of Geographical Databases' 243–56 © 1994, with permission from Elsevier Science.

- each point has no null co-ordinates
- each edge has two different nodes, and not other points
- each edge has two faces
- each face is represented by a simple circuit
- each circuit must be closed
- no intersecting faces
- no node within a face
- the corresponding graph is connex.

Based on those constraints, Ubeda (Figure 4.24) defines a topological integrity constraint by:

```
(Geo-object class 1, topological relation, Geo-
object class 2, specification)
```

By geo-object class, one can give any geographic features such as streets, buildings, lakes and so on. The topological relation are based on the 9-intersection model proposed by Egenhofer (1991). So an example can be:

```
(Road, cross, building, forbidden).
```

In general the constraint must hold for all roads and buildings. But, when in a city, there are a few cases where this constraint does not hold. A relaxing mechanism can be put in place to prevent finding an error for those special situations.

Starting from the characteristics of geographic objects, Ubeda (1997) has established the following properties for geometric objects (Figure 4.25):

- p1:   consistency of points
- p2:   unicity of punctual elements (points and nodes)
- p3:   unicity of linear elements (segments, lines and arcs)
- p4:   unicity of surface elements (polygons and nodes)
- p5-1:   openness of lines
- p5-2:   closure of polygons and faces
- p6:   non-intersecting feature

(Geo-object class1, Topological relation, Geo-object class2, Specification)

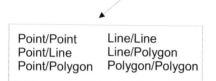

Point/Point	Line/Line
Point/Line	Line/Polygon
Point/Polygon	Polygon/Polygon

Forbidden
At least n times
At most n times
Exactly n times

Figure 4.24 Definition of topological integrity constraints.
According to Ubeda (1997).

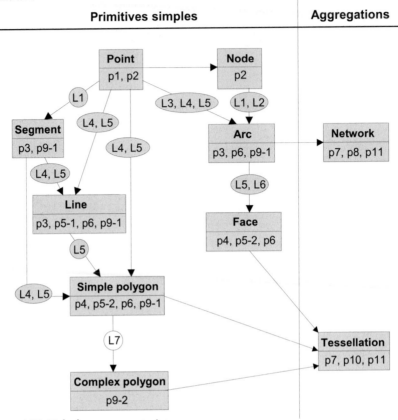

Figure 4.25 Links between constraints.
After Ubeda (1997).

- p7: connexity of the feature
- p8: good-looking networks
- p9-1: orientation
- p9-2: orientation of the contour
- p10: total coverage of the space
- p11: non-overlapping of elements within a tesselation.

And also the properties between geographic objects:

- L0: referential integrity (not mentioned in Figure 4.25, but valid everywhere)
- L1: the two points or nodes are different
- L2: belonging at least to two objects
- L3: belonging to exactly one object
- L4: all objects within a list are different
- L5: the list objects are ordered

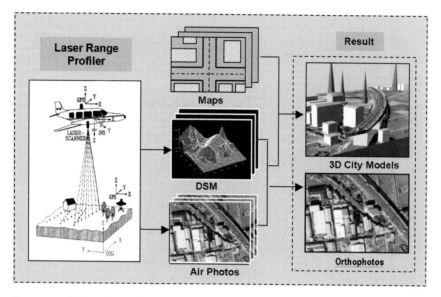

Figure 4.26 The outline of a system for generating a 3D model of a city based on a range profiler, digital terrain models (DSM) and aerial photos.
From Chen, X. 1999 used with kind permission.

- L6: belonging to exactly two objects
- L7: belonging at most to two objects.

Then the database is scanned to check whether all objects follow the corresponding topological integrity constraints. The detected errors are then stored into a file which is examined. For each error, some correction scenarios are proposed, and the user must select one of them.

4.9 Example of multi-source updating: mixing aerial photos and laser data

As an example of multi-source updating let us review very rapidly the example given by Chen X. (1999) based on an airborne range profiler, a terrain model and aerial photos in order to build a 3D model of the city as depicted Figure 4.26.

4.10 Conclusions

The goal of this chapter was to give the reader some elements for quality control and assessment in urban databases, specially insisting on the necessity to have a maintenance plan and regular updating. But updating geographical databases, especially by using multi-source integration is a very complex problem. Firstly it was shown that updating can have different meanings and the conventional way of updating relational databases (by means of SQL or the conventional way of updating relational databases (by means of SQL or PL/SQL) is not effective in this field. Indeed, integration of multi-source information is a more frequent case.

More investigations are needed in order to confront several sources and use them for updating. As explained, in some cases, automatic updatings are not possible and in this case, visual interfaces must be provided to the user. In addition, when automatic updates are possible, more efficient algorithms must be created. Anyway, local authorities must implement a good data quality maintenance plan (see Roberts (1999) for details). In Chapter 6, the problem of metadata will be dealt with, in which quality information will be integrated.

To conclude this chapter, let us say some words regarding the legislative aspects. Indeed, some parcel co-ordinates are generally measured by surveyors and those measures can have juridical consequences. Under those circumstances, perhaps updating a database must imply the validation by some lawyers. Of course, this aspect is beyond the scope of this book, but people working in geoprocessing must be aware about the importance of this fact. So, the design of updating interfaces is a very nice solution allowing the user to control the way of performing this task.

5

HYPERMAPS AND WEB SITES FOR URBAN PLANNING

Map-based hyperdocuments, hypermaps, clickable maps are different names for new tools for organising spatial information and documentation using the Internet. Among the possibilities, let us mention navigation and the design of those new kinds of documents for urban planning purposes. Examples will be taken from tourism and road maintenance applications, although an example for land-use planning will be described later in the groupware chapter (Chapter 8). But urban planners generally in their office keep not only information on the territory they have to plan, but also on experiences in different cities all over the world: 'out-of-my-backyard' hypermap structuring could be an interesting tool for organising this particular documentation. The Internet as a very important network allows us not only to access simple hyperdocuments, but more importantly bases of hyperdocuments with transactions. Under the name of Intranets, such systems can be now commonly found.

In this chapter we move closer to data organisation for a higher semantic level. Here, we deal with data of different media, for example photographs and sound, as well as text and graphic materials that, ideally, we would like to store and access in the same digital spatial information system environment. The organisation of such diverse data through the hyperdocument approach is then presented as the specific topic of hypermaps. Then considerations regarding web sites will finish this chapter.

5.1 Hypertexts and hyperdocuments[1]

Hypertexts and multimedia hyperdocuments are an increasingly common type of documentation. Whereas most, but not all, conventional documents are essentially textual print media, have a logical and a physical layout structure

---

1   A preliminary version of this chapter can be seen in Laurini and Thompson *Fundamentals of Spatial Information Systems*, Academic Press, 1992, Chapter 16 which was based on the reference paper Laurini and Milleret-Raffort (1990).

organised sequentially and hierarchically, hyperdocuments are a modern version of non-linearly organised materials. That is, they are electronic documents with direct access to information of diverse form by means of window presentation and mouse clicking on important words or other displayed information.

There are two principal facets to the domain of spatial data beyond the realm of textual information in print media:

- the non-print media
- the non-sequential organisation and access.

The non-print media, often referred to as multimedia in the jargon of the computer and electronics industries, comprise a variety of analogue and digital forms of data that come together via common channels of communication. Traditional audio-visual and text forms are assembled electronically but the net can be cast wider to include seismic signals, sound, and even (in the future, perhaps) data produced via other human sensations, tactile or olfactory. We use the term content portion to refer to the individual unit for data, of whatever form. The non-sequential organisation, or hyperstructure refers to a form of communication beyond or over the linear style that is associated with most books.

5.1.1   Multimedia spatial data

In libraries the non-print media collection comprises picture material in the form of photographs, filmstrips, 35 mm slides, videotapes or videodisks. Some printed materials may include or consist entirely of monotone or colour photographs. Many scientific fields have long used sketches, maps, graphs, charts, diagrams or other non-textual materials. For more than thirty years many data have been remotely acquired in electronic form via special instruments recording data from such phenomena as traffic or air pollution, and in digital form from spacecraft-borne sensors. Dealing with varied information is not a new dimension to many fields but being able to integrate analogue and digital forms, and to work with print and non-print media, continue to challenge the electronic engineers and software designers.

In the context of urban information systems there are many uses of non-print media, and some interesting applications blending the world of pictures, sounds and signals with the conventional text- and map-oriented database software and hardware. Among the possible domains of use of multimedia data in urban information systems are the following:

- urban and regional planning
- environmental planning
- hazard prevention and management
- fire fighting

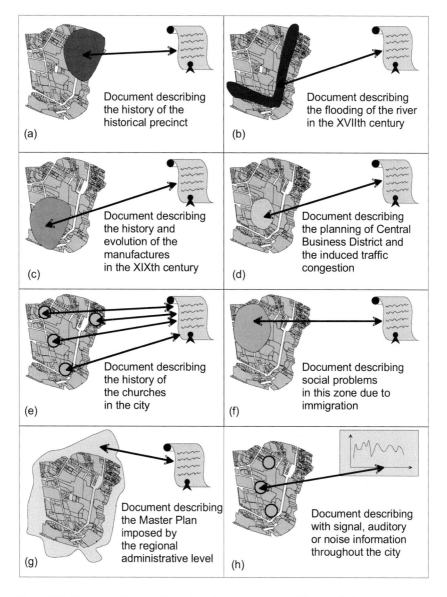

Figure 5.1 Example of co-ordinate-based documents. (a) Historical documents. (b) Documents concerning floods. (c) Documents relative to the history of manufacturing. (d) Study regarding the impact of the Central Business District to traffic conditions. (e) Document describing the architecture of some churches. (f) Study describing some social problems due to immigration. (g) Connections with the Regional Master Plan. (h) Document describing auditory or noise information with signals.

- tourist industry
- road maintenance
- historical garden and sites
- history of the city
- engineering networks
- information for investors
- new home seeking.

But we have more than a combined use of multimedia data; we have also multimedia documents for which a geographic access is most important. By geographic access, we mean a co-ordinate-based access in which by referencing a place or a region on a map, we can retrieve all information dealing with the identified point or zone. Referring to Figure 5.1, we need, for example, to demarcate a region by any arbitrary boundary defined by a random computer mouse clicking on a map image, as a way to retrieve not only numerical or alphanumerical data, but also documents, images, measurements and so on, 'grounded' or anchored at some point or in some region.

At the moment we often have huge collections of paper documents serving many purposes. These consist of maps, architectural drawings, engineering blueprints, land-use zoning maps, sketches and the like for which an access based on co-ordinates is quite desirable. A typical query can be to retrieve all documents describing the totality or a portion of a territory defined by mouse-clicking on a point or on the boundaries of a zone. Consequently we need to store, to archive, to select and to retrieve materials such as maps with different scales and topics, sketches with different approximation degrees, paper reports and documents with various origins and structures, photographs and three-dimensional landscape images with different geometric perspectives, aerial photographs and satellite images with unknown exact positions, seismic signals, temperature measurements and all sorts of electromagnetic signals.

5.1.2   The hypertext concept

Traditional texts, linear or hierarchical in form, have a double structure; logical, comprising chapters, sections, paragraphs and so on, and layout, consisting of volumes, pages, blocks and so on. In contrast, hypertexts are organised by semantic units called nodes and associations between nodes referred to as links. The semantic units may be simple elements like a single word or proper name, or may be more complex like a table of numbers, or a map, or photograph, or a paragraph of words describing an intellectual concept or an entire document or file of data. Most users of hypertext favour using nodes which express a single concept or idea. A hypertext system invites the creators to modulate their own thoughts into units in a way that allows an individual idea to be referenced elsewhere. A node may have several successors and predecessors; readers can choose their own path of reading.

Generally, a hypertext possesses a great many short semantic units and every unit is connected to others by means of reference links, sometimes named anchors. Links can be of several types, such as:

- to connect a node reference to the node itself
- to connect an annotation or source citation to a document portion
- to provide relationships between two objects within the same document
- to connect two successive document portions.

There is in this network of nodes a large quantity of information in separate fragments which can be linked in different ways, and for which only a few items may be used at only one time. Linked by semantic associations, the set of information is inherently a loose, unbounded structure something like the pages of a thematic atlas, with unlimited possibilities of connection, but with some tighter bindings, like an inset map for a smaller-scale map. Knowledge of the substantive domain can reveal patterns of connection that make more sense than others, so that we create particular graph instances called webs, representative of some order. With this form of linkage a hypertext has a network structure. When the number of references among the semantic units is high, this structure becomes a cobweb of links. Figure 5.2 depicts a cobweb of hyperdocuments in which nodes appear as squares, a reference location as a dash and links as unbroken lines.

In an electronic environment the connecting is done at two levels. Icons, windows or other forms on a screen represent entities in a database and can be linked by graphical means on the screen, while in the database there are connections by pointers or some other device. For navigating from one node to another, we have to click a reference word or picture which constitutes a bridge to another semantic unit. So, by a man-machine interface, especially based on windows and a mouse. it is possible to move from one semantic unit

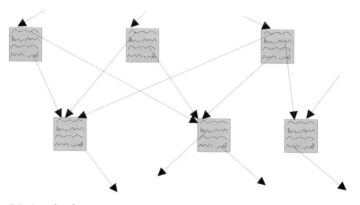

Figure 5.2 A web of semantic units.
From Laurini and Thompson (1992).

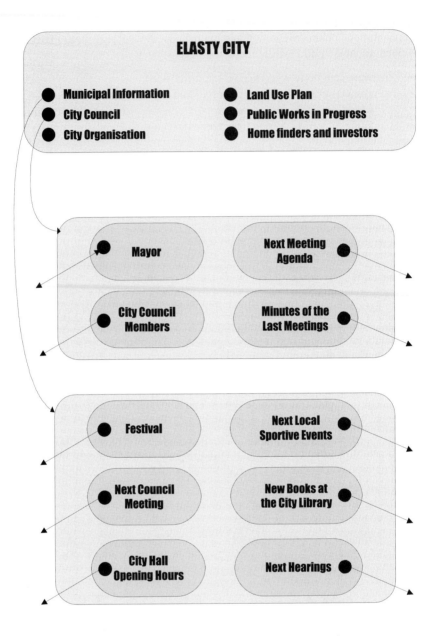

Figure 5.3 An example of a hyperdocument network for a municipality.

to another. This provides a way to 'read' electronically a dossier about some topics, regrouping several small documents together and it provides a means for learners to make associative linkages between apparently unconnected, discrete pieces of information. This hyperdocument concept is far from the conventional printed book with a fixed hierarchical structure.

The same structure can be built not only for text material (hypertext), but also for multimedia contents such as images, graphics, sounds and signals. In this case, we deal with hyperdocuments and hypermedia. Typically, browsing a hyperdocument is done via several mouse clicks as for dictionaries in which the reader is going from one word to another. We can by this approach intertwine the main text and footnotes, figures, tables and images as suggested by Figure 5.3 which gives the entry point of a hyperdocument presenting some local information for city dwellers.

The main realisations of the hyper concepts are:

- A hypertext is a network of content portions which are always textual.
- A hyperdocument is a network of content portions which can always be displayed on a screen or presented via a loudspeaker.
- In hypermedia, the content portions are like the hyperdocument but can also include digitised speech audio recordings, movie film clips and presumably tastes, odours and tactile sensations.

These are the main concepts used to construct the now conventional Internet web site. Let us use the term hypermap to refer to multimedia hyperdocuments with a geographic co-ordinate base accessed via mouse clicking or its equivalent. It is the objective of this chapter to define exactly the principles of hypermaps, their applications, and give some highlights on their physical structures.

5.1.3   Portals

Portal is a new term, generally synonymous with gateway, for a World Wide Web site that is or proposes to be a major starting site for users when they get connected to the Web or that users tend to visit as an anchor site. A number of large access providers offer portals to the Web for their own users. Most portals have adopted a style of content categories with a text-intensive, faster loading page that visitors will find easy to use and to return to. Companies with portal sites have attracted much stock market investor interest because portals are viewed as able to command large audiences and numbers of advertising viewers. Typical services offered by portal sites include a directory of Web sites, a facility to search for other sites, news, weather information, e-mail, stock quotes, phone and map information, and sometimes a community forum. The term portal space is used to mean the total number of major sites competing to be one of the portals.

A well presented front-page must be designed as a portal, allowing the user to offer as many possibilities as possible. Recently, Peinel and Rose (1999) have proposed a geographic portal which is built on the concept of smart maps for information visualisation and exploration. These smart maps serve as a navigational hub by indicating the existence of information carrying relevance for a specific location. Groupware functions for collaborative visualisation and interactive editing on maps are added on top in order to export the concept of graphical information portals to further domains, such as risk management. Figure 5.3 can be seen as a portal for the municipality.

5.2 Hypermaps

The special idea of hypermaps (Laurini and Milleret-Raffort 1991, Kraak and van Driel 1997) is to extend the hyperdocument concepts by integrating geographic referencing. Some authors now use the expression 'clickable maps'. Should geographic references remain at a literal level, such as by place names, they can be modelled by hypertext links. For many documents, though, this kind of reference is not enough to cover correctly the space involved; a co-ordinate-based referencing method through a cartographic system is desirable. We now examine the implications of co-ordinate-based documents according to two features; spatial referencing and spatial queries.

5.2.1 Spatial referencing of hyperdocuments

Spatial referencing of hyperdocuments presents two aspects (Figure 5.4):

* spatial referencing of document nodes
* spatial referencing of maps and other cartographic documents.

A node, supposedly representing a single or a few semantic concepts, may have one or many spatial references. For example, a particular document node can describe a geodetic point, a region or a linear feature (Figure 5.5). Some other nodes can compare points and regions; thus, we can have multi-point document nodes, multi-line document nodes, multi-area document nodes, and several other combinations.

Let us take the conventional way of dividing spaces into zero-dimensional elementary objects (for point), one-dimensional (for line), two-dimensional (for area) and three-dimensional (for volume) entities, as in Figure 5.6. As documents can be associated to any elementary object, here referred to as zones, elementary objects can also be associated to documents. A sort of many-to-many relationship can then be defined. In order not to have problems with elementary objects, let us suppose that spatial referencing is made through elementary pieces of space, via a tessellation. For the method of organising and indexing this kind of information, please refer to Laurini and Thompson (1992).

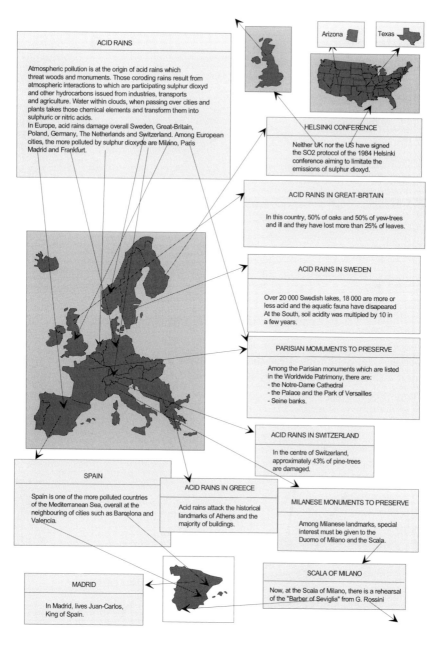

ACID RAINS

Atmospheric pollution is at the origin of acid rains which threat woods and monuments. Those coroding rains result from atmospheric interactions to which are participating sulphur dioxyd and other hydrocarbons issued from industries, transports and agriculture. Water within clouds, when passing over cities and plants takes those chemical elements and transform them into sulphuric or nitric acids.
In Europe, acid rains damage overall Sweden, Great-Britain, Poland, Germany, The Netherlands and Switzerland. Among European cities, the more polluted by sulphur dioxyde are Milano, Paris Madrid and Frankfurt.

Arizona

Texas

HELSINKI CONFERENCE

Neither UK nor the US have signed the SO2 protocol of the 1984 Helsinki conference aiming to limitate the emissions of sulphur dioxyd.

ACID RAINS IN GREAT-BRITAIN

In this country, 50% of oaks and 50% of yew-trees and ill and they have lost more than 25% of leaves.

ACID RAINS IN SWEDEN

Over 20 000 Swedish lakes, 18 000 are more or less acid and the aquatic fauna have disapeared At the South, soil acidity was multiplied by 10 in a few years.

PARISIAN MOMUMENTS TO PRESERVE

Among the Parisian monuments which are listed in the Worldwide Patrimony, there are:
- the Notre-Dame Cathedral
- the Palace and the Park of Versailles
- Seine banks.

ACID RAINS IN SWITZERLAND

In the centre of Switzerland, approximately 43% of pine-trees are damaged.

SPAIN

Spain is one of the more polluted countries of the Mediterranean Sea, overall at the neighbouring of cities such as Barcelona and Valencia.

ACID RAINS IN GREECE

Acid rains attack the historical landmarks of Athens and the majority of buildings.

MILANESE MONUMENTS TO PRESERVE

Among Milanese landmarks, special interest must be given to the Duomo of Milano and the Scala.

MADRID

In Madrid, lives Juan-Carlos, King of Spain.

SCALA OF MILANO

Now, at the Scala of Milano, there is a rehearsal of the "Barber of Seviglia" from G. Rossini

Figure 5.4. Example of a hypermap.

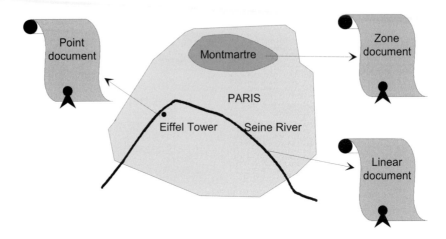

Figure 5.5 Documents linked with various types of geographic zones, namely points, lines and zones.

5.2.2 Spatial queries for retrieving hypermap nodes

To retrieve a spatial document, the basic starting point is a map against which we query by delimiting a region by means of a mouse. Four types of spatial query are of relevance: point, buffer zone, segment, and region query. So, navigating in a hypermap must combine these types of query and conventional hyperdocument scanning via reference links (Figure 5.6).

When making a hypermap query, several situations can occur (Figure 5.7). We can see that documents *D2* and *D3* are fully relevant, document *D1* is

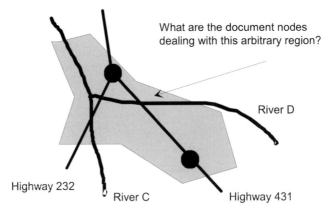

Figure 5.6 A hypermap region query.
From Laurini and Thompson (1992).

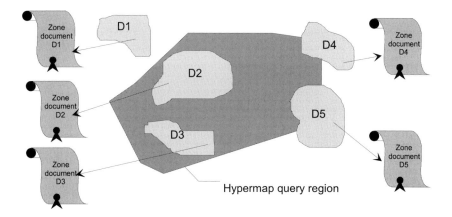

Figure 5.7 Relevant, partially relevant, and irrelevant documents.

irrelevant, and documents *D4* and *D5* are partially relevant. Sometimes, it is difficult to have the exact demarcation zone for a document node, in which case a membership degree can be given to several embedded zones. For details, please refer to Laurini and Thompson (1992) about fuzzy concepts. In the example given in Figure 5.7, notice that document *D2* has for instance 100% fit, but *D4* is only about 20%. However, the main challenge is how to organise both locational relationships and document relationships. We have two aspects to consider:

- the document-to-map relationships
- the map-to-map relationships.

5.2.3 Active zones[2]

In order to access documents from co-ordinates, two possibilities exist depending of the type of active zones. According to the type of information we want to access, either all the map points are active, or some special locations are active. The former situation refers to a continuum of possibilities to access documents, whereas in the latter, only a few zones, generally round or squared, are active, appearing as anchors to documents. With this method, a territory or a map is not totally covered by documents, but only a few active zones exist as illustrated in Figure 5.8 showing a city-ward with only four active zones in black. In other words, active zones are buttons which we can press to get information, whereas in the other solutions, the point-in-polygon query is run.

---

2   See an example for the city of Oxford, UK at: http://www.oxfordcity.co.uk/maps/ox.html.

Figure 5.8 This map shows only four active zones represented as small squares used as entry points to documents.

5.2.4 Mobile anchors for images and movies

Active zones can be put on images easily; for instance, in a group of people, when clicking on somebody, one can get information. For movies, a similar system can be implemented. For instance Figure 5.9 is extracted from a movie made by some researchers at the University of Tokyo (Murao *et al*. 1999) showing the name of a building (an hotel in the example) and its URL, which are acting as active zones in order to visit the web page of this hotel. Of course, during the evolution of the movie, the name and the URL are moving in order to be still written on the building, so getting mobile anchors. These URL can be written directly on images. But another possibility is to situate comments with a virtual device as proposed by Höllerer *et al*. (1999) in order to embed multimedia presentations into the real world.

5.3 Navigation in hypermaps

Navigating in a hypermap features two aspects: **thematic navigation**, as is usually done in hypertext and in hypermedia, and **spatial navigation** which is particular to hypermaps. By thematic navigation we mean the way to navigate from a text to something else. In order to handle several kinds of user several modes of navigation can be implemented. Thus we can define a novice mode

Figure 5.9 Example of URL anchored on building within a movie (Murao *et al.* 1999). Used with permission of Michihiro Murao.

when the user is a newcomer, and an expert mode when the user has a good background in the domain.

As an example of a human-machine interface emphasising both the spatial and the thematic aspects of hypermap navigation, we can provide a context as to how this environment might be. The map is the window to the world, the entry point to the hypermap system, the access to a multimedia resource. Imagine a tourist visiting New York City, browsing through a map showing the main attractions by pictorial icons. A click or touch on a symbol for the Statue of Liberty provides a connection not only to a picture of that gift from the people of France, but also a connection to a different part of the world, possibly to retrieve some textual information describing the context of that donation. For another example regarding cities, visit the site of Chicago, http://www.cityofchicago.org/Tourism/Downtown/index.html.

5.4 Designing hypermedia

For the design of hypermedia or an internet web site, there exist several methodologies. In this section, we will introduce only the RMM and some recommendations.

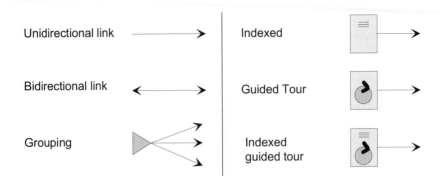

Figure 5.10 The RMM primitives which are added to the entity-relationship model.

5.4.1 Example of methodology for hypermedia design

Relationship Management Methodology (RMM) (see Isakowitz *et al*. 1995) can be seen as an extension of the entity-relationship methodology (see Section 2.3) for the design of hypermedia. In addition to hypermedia entities and attributes, several types of links can be distinguished as given in Figure 5.10. By unidirectional link, a single pointer is used, by bi-directional links, two pointers are used for the way forth and back.[3] Grouping represents the cases in which several kinds of entities must be accessed from this point. Two new concepts must be presented, index and guided tour. By index, we mean that a list of entities can be accessed through a list, and by guided tour, the entities can be accessed through a ring data structure. Index and guided tours can be combined to give indexed guided tours as illustrated by examples in Figure 5.11.

Global RMM methodology is presented in Figure 5.12. The first step is the requirement analysis. This step is very difficult to implement because lay users generally do not know what they want, especially when dealing with new technologies. A nice idea is to start with a prototype which is evaluated and corrected. The next stage is the design of the diagram with the tools previously introduced, not only for the selection of entities, but also for the type of accesses for navigation. After that, there is the design of the user-interface and especially the screen design. After having chosen an authoring system, the real construction of the hypermedia can begin.

5.4.2 Some recommendations

Recently Lynch and Horton (1997) gave some recommendations for hyper-document design. The analysis is that 'loose links can drive away an audience, dilute the site's message, confuse the reader with irrelevant digressions, and

---

3   Bidirectional links are generally not implemented.

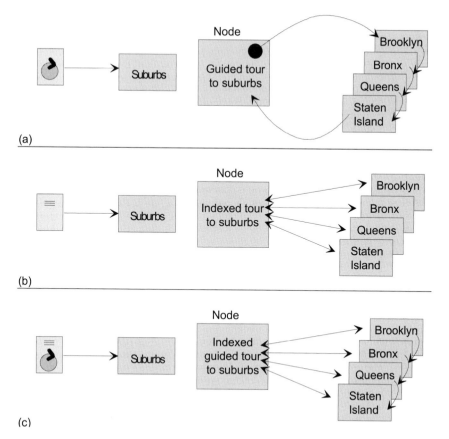

Figure 5.11 Examples illustrating ways to access to hyperdocuments. (a) Guided tour. (b) Indexed tour. (c) Indexed guided tour.

become a continuing maintenance headache for site authors'. They say that at the moment, hyperdocument pages resemble more tables of contents and resource bibliographies and they cannot replace real commentary, analysis, or sustained rational arguments. Among guidelines for effective linking, they suggest:

- Links should reinforce messages, not replace them.
- Most links should point to your site, not away from it.
- Most links should appear as footnotes, away from the main text.
- Links to outside sites should open a new browser window.
- Every link is a maintenance issue; link sparingly, if at all.

5.5 Some examples of hypermedia design

In order to illustrate the way to design hypermedia, let us examine some examples taken from different domains, which will be detailed in this section:

137

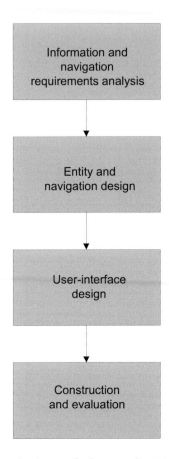

Figure 5.12 The main steps in the design of a hypermedia (RMM).

- public services
- land-use planning
- road maintenance
- for home seeking and investors
- out-of-my-backyard
- and finally for teaching, urban world.

5.5.1 Public services in cities

For a city, it can be interesting to organise information targeted to citizens, relative to public services located in its territory with hyperdocuments. The portal can be either a map or a space-filling treemap (see Section 7.3.1) of public services. Figure 5.13 gives a structure able to drive the readers, for instance to schools, medical centres, social (welfare) centres, libraries, sport clubs, and so on.

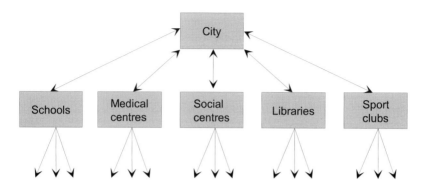

Figure 5.13 Portal of a hypermedia for municipal public services.

5.5.2   Hyperdocuments for land-use planning

Before designing a new house or some commercial premises, an applicant might be interested to know what are the building regulations in order to know whether a 10-storey office building can be constructed, or whether he can extend his house or make a camping site. For this purpose, a hypermap can be an interesting tool giving for each zone, not only a prescriptive land-use map, but also written regulations as illustrated in Figure 5.14. The hypermap structure is given in Figure 5.14a. For acquiring a copy of the regulations, the applicant can fill a form in order:

- either to download the concerning prescriptive land-use plan and the written regulations
- or to receive a copy by means of postal or electronic mail.

5.5.3   Road maintenance

For road maintenance, it can also be interesting to create a hypermap. Suppose a Road Department within some local authority wants to organise all information regarding road maintenance. This information can take several formats, reports, maps, plans, estimates, invoices, traffic statistics, photographs and so on. In addition, this information is not only located in space, but also corresponds to special dates. As exemplified in Figure 5.15, all those documents can be anchored to some locations in order to set up a hypermap system which is sketched in Figure 5.16.

5.5.4   For home seeking and investors

People looking for new homes and investors are a new class of users of urban information. Web documentation can help them a lot. For new home seekers, it

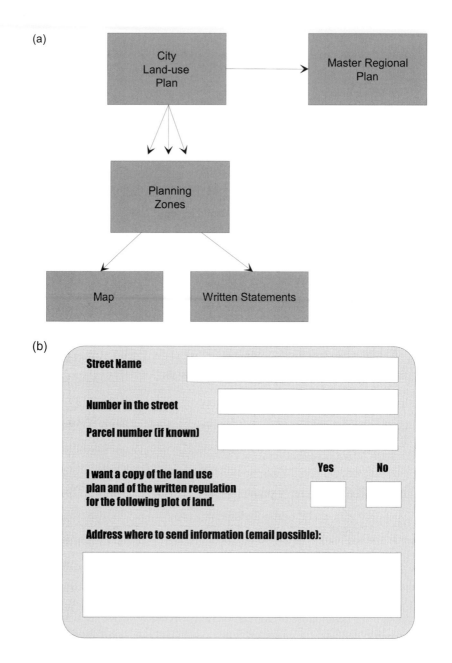

Figure 5.14 Hyperdocuments and hypermaps for land-use planning together with a form to fill in to receive by post a copy of the regulations. (a) Hypermap structure. (b) Form to fill in.

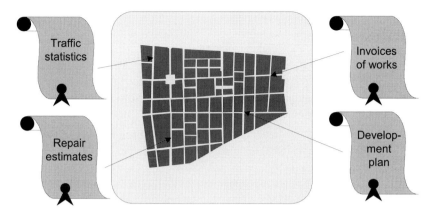

Figure 5.15 Example of a hypermap system for road maintenance.

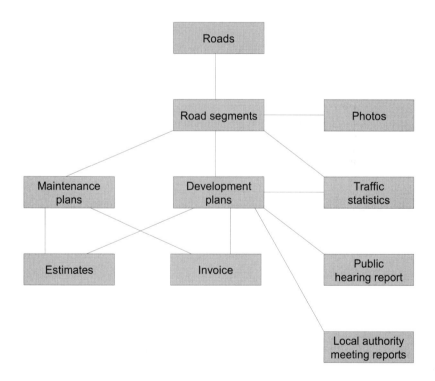

Figure 5.16 Structure of a hypermap for road maintenance.

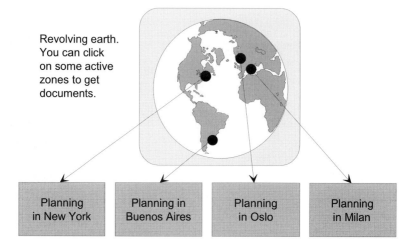

Figure 5.17 'Out-of-my-backyard' hypermap for urban planning.

is interesting not only to have a photo of the house, but also to have information about the neighbourhood, for instance concerning local public services, schools, shops, churches, transportation, crime information and so on. For companies seeking to relocate or to open a new branch, all information regarding the local business marketplace will be very useful. For example, the reader can refer to http://wwwcityofchicago.org, or http://www.cityofseattle.net.

5.5.5 'Out-of-my-backyard' hypermap

Urban planners not only keep information about the city they have to plan, but also information of planning experiences throughout the world, which can be of interest for them. Similarly, this information can be organised as a hypermap. Under the name of 'out-of-my-backyard' hypermap, any urban planner can organise this kind of information. Figure 5.17 gives an idea of the entry point. It can be a conventional 2D map of the earth, or a revolving earth as illustrated. Some active zones are located throughout the world as anchors to documents, maps and so on. But when the earth is moving, clicking on the fly, some hypermap anchors perhaps need the agility of synchronising movement of the hand and eye.

5.5.6 UrbanWorld (Thompson et al. 1997)

The UrbanWorld Hypermap Learning Environment Project was established to evaluate the use of geographic information system technologies and digital data sources in current urban geography and planning education. The focus of this effort was to establish a series of learning modules that contain a variety of subjects complete with data sources for students to explore and analyse.

The UrbanWorld project, developed at the University of Maryland Department of Geography, was a three-year project that began in January 1996. Funding was provided by the US Department of Education Fund for the Improvement for Post-Secondary Education (FIPSE) Additional support came from ESRI, Inc. (Environmental Systems Research Institute, Redlands California). For details, refer to: http://www.inform.umd.edu/geog/gis/urbanworld/html/about.html.

5.6 Intranets and extranets

According to Turban and Aronson (1998), an intranet is a network architecture designed to serve the internal informational needs of an organisation using the Internet concepts and tools. It provides similar capabilities, namely easy and inexpensive browsing communication and collaboration. Under a consistent typical user interface applications include:

- publishing corporate documents
- providing access to searchable directories such as telephone and address lists
- publishing corporate, departmental, and individual pages
- providing access to groupware applications (see Chapter 8)
- distributing of software
- providing e-mails.

An intranet is relatively safe within the organisation's firewalls. A firewall is a method of isolating the computers behind a device that acts as a gatekeeper for the whole organisation.

As an intranet is targeted at users within an organisation, an extranet is devoted to users outside an organisation, but tightly linked with the organisation. For companies, extranet users are customers and providers. For a city, an extranet can be targeted at citizens, or any person looking for information regarding the city such as tourists, investors, etc. As an example, let us examine the extranet of the Italian city of Turin for urban planning, and some elements for on-line permitting.

5.6.1 Extranet of the city of Turin (Gauna and Sozza 1999)

The city of Turin, Italy has recently developed an extranet concerning its land-use plan. So, each citizen can know the regulations which apply to his parcels. Moreover, when somebody, or a company wants to construct a new building, they immediately know the planning regulations of the zone. This corresponds to the first step in making preliminary studies. One must proceed as follows:[4]

---

4 http://sit.comune.torino.it/SIT.html

Figure 5.18 Portal of the city of Turin.
Used with kind permission of Comune di Torino.

1   Display the main page of the Turin GIS portal; on the left of the screen, one can see all the services which are accessible, albeit in Italian (Figure 5.18). Choose the cartography sub-system and a new page is opened (Figure 5.19).

2   Select the subject: in the new page it is possible to choose between the topographic map or the Master Plan. By choosing the Plan option access to the system of the city planning scheme is made (Figure 5.20).

3   Consult the Land-use Master Plan, against which some hypermap-style query can be made. The query starts from one of the grid cells that form

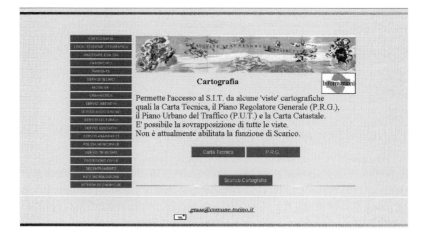

Figure 5.19 Accessing cartography services.
Used with kind permission of Comune di Torino.

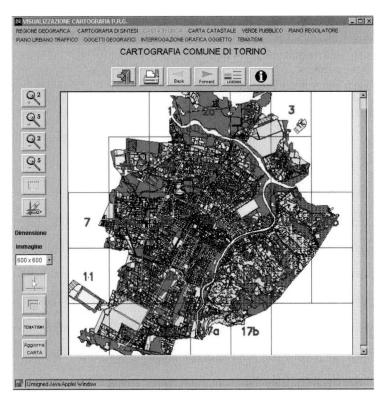

Figure 5.20 The Turin land-use plan (Piano Regolatore Generale).
Used with kind permission of Comune di Torino.

the general map. By choosing a particular area it is possible to obtain information on all the objects that are touched in the query. The result of the query is visualised in a table (Figure 5.21).

4   Obtain the final output of the query. Every one of the seven folders corresponds to a class of objects and summarises the data of the single object touched with the query belonging to that class.

5.6.2   On-line permitting

Thanks to this new technology, such new applications concerning urban planning can be envisaged, one of the hot issues being online permitting. Issuing a building permit or any kind of authorisation can now be done via the Web. In this case, one speaks about digital submission of documents. Figure 5.22 depicts an interface for online permitting. The user can:

• submit a new permit
• give his comments regarding the permits presently under study

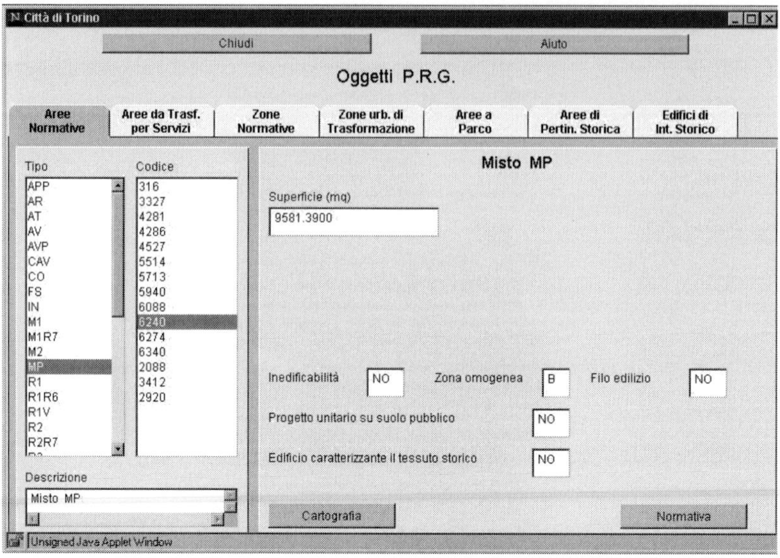

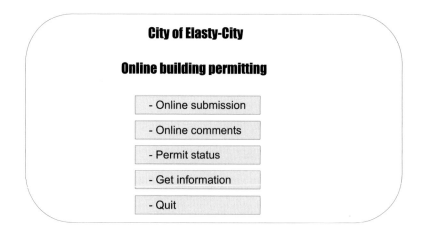

Figure 5.21 Example of a query result.
Used with kind permission of Comune di Torino.

- know about the status of his own submission
- get information regarding the system and the legislative context.

See an example in Indianapolis, USA, http://www.ci.indianapolis.in.us/dmd/ permits/ or in West Virginia http://www.dep.state.wv.us/permit/pi_ mgmt.html.

City of Elasty-City

Online building permitting

- Online submission

- Online comments

- Permit status

- Get information

- Quit

Figure 5.22 Example of front-page on line permitting.

5.7 Summary

In this chapter we have defined hyperdocuments and hypermaps for their use in urban planning. During the last few years, interesting systems have been developed in cities either to inform citizens, or to facilitate the tasks of urban planners. For the next few decades, I think that very important progress will be made especially towards the connections between the local administrations and citizens. Some people speak now about cyber-citizens, or e-citizens (see Chapter 9). But the main problem in defining a web site is still to answer correctly three questions:

1 Who will the users be, and what do they look for?
2 What kind of services to provide?
3 What kind of image does the organisation want to provide?

In other words, the key issue is the purpose of the web site to be designed.

6

FROM URBAN KNOWLEDGE
TO SPATIAL
METAINFORMATION

The scope of this chapter will be to study knowledge and metainformation for urban planning. Remember, as suggested in the second chapter, first we have data, then from data we can create information and eventually knowledge.

Looking back, we can see an evolution in the meaning of these words. Indeed, during the 1980s, that is to say during the expert system years, 'knowledge' was often kept as a synonym of 'rule'. Facing a new application, the key-idea was to capture knowledge, or to extract some knowledge bunch by discussion with human experts. These were perhaps the 'golden days' of artificial intelligence. Now, the expression 'artificial intelligence' is less used, and the expression 'knowledge engineering' (Debenham 1998) is more common, with a broader and a more flexible sense of knowledge.

In parallel with this evolution, it has appeared more and more important to describe data correctly, or to give them precise information concerning the definition, its value: acquiring and structuring metadata together with data is presently very common. Further, I think that nobody today will buy data without requiring also metadata, generally by exchanging or purchasing two files, data items themselves and metadata. For the first step, we will point out some generalities regarding these issues, stressing the importance of the 'meta' side, and a new approach based on ontologies. Then all aspects regarding knowledge will be defined in order to present expert systems and knowledge base structuring. After, argumentation will be presented as a special case for presenting opinions regarding urban or environmental planning disputes and the way of solving those disputes. We will finish this chapter by presenting spatial metadata and their importance in structuring information systems.

6.1 Generalities

Accepting 'information' in the broader sense, we can distinguish three levels, data, information (narrow meaning), and corresponding to these three levels (see Figure 6.1), three 'meta' levels, namely, metadata, metainformation and metaknowledge. The interest in using multiple levels of abstraction

Levels of abstraction

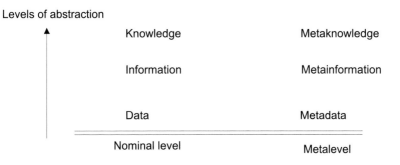

Figure 6.1 Scale of information levels, scale of levels of information about information.

corresponds to the necessity of conceptualising what we know about the world in order to easily reason about it.

According to Debenham (1998), 'given an application, a data thing is a fundamental indivisible thing in this application. An information thing is an implicit association between data things. A knowledge thing is an explicit association between data things or information things.' For him, an explicit association can be described by a succinct, computable rule, whereas an implicit association cannot be described succinctly and computationally.

6.2 Knowledge[1]

In Table 6.1 according to Turban and Aronson (1998 p. 205), some issues, emphasising the interest of using knowledge instead of data, or more exactly, the potentialities of programming based on knowledge versus conventional programming is presented. Perhaps the main advantage is to state that knowledge is independent from the application, and also may be incomplete or not yet completed. One of the definitions of knowledge is to say that knowledge can be decomposed into facts and rules (Laurini and Thompson 1992).

Facts can be defined as single events, such as the vehicle collision that occurred at the intersection of First Street and Broad Avenue, or single features, for example the value of an attribute or an entity. Other examples are: is it raining? Peter is three years old, the Eiffel Tower is 300 metres high, Moscow is the capital of Russia.

Rules can be many kinds of statement that establish a regulation, a process, a method, a standard, a principle, a code of conduct, a law, a procedure, an ordnance. Generally speaking, they are written with IF-THEN statements (sometimes called rules) such as:

1   A subset of the section was issued from Laurini and Thompson 1992, Section 17.5.

Table 6.1 Knowledge versus data engineering. From Turban and Aronson 1998, p. 205

Dimension	Based on knowledge	Based on data
Processing	Primarily symbolic	Primarily algorithmic
Nature of input	Can be incomplete	Must be complete
Search	Heuristic (mostly)	Algorithms
Explanation	Provided	Usually not provided
Major interest	Knowledge	Data, information
Structure	Separation of control from information (data)	Control integrated with information (data)
Nature of output	Can be incomplete	Must be correct
Maintenance and update	Easy because of modularity	Usually difficult
Reasoning capability	Limited but improving	None

IF a set of conditions occurs THEN a set of actions follows

or with a mathematical expression:

IF C_1 and C_2 and . . . and C_l THEN A_1 and A_2 and . . . and A_m

for which the semantic context can be stated as:

IF a context is realised THEN do so and so.

Sometimes, instead of conditions, the word premise is used. For example:

IF it rains THEN take an umbrella

In this trivial example we have two specific facts: 'it rains', and 'take an umbrella', linked by a rule when the first fact, the condition, occurs. Now let us examine other examples more germane to the scope of this book. In strategic urban planning, some examples of rules might be:

IF a zone is a marshland THEN prohibit construction

or

IF there is a high level of unemployment
THEN encourage business enterprise creation
AND create industrial areas

or

IF a parcel is close to an airport
THEN limit building height

or

IF a parcel is near to a fire station
THEN prohibit hospital construction.

In urban renewal:

> IF a building has a good architectural design
> AND IF it is more than 100 years old
> AND IF the building condition is mediocre
> AND IF the owner agrees
> THEN suggest restoration

or

> IF a building falls into ruin
> AND IF nobody dwells in it
> THEN demolish it

or

> IF a building has a poor architectural design
> AND IF inner rooms are deteriorating
> AND IF money is raised
> THEN suggest rehabilitation

6.3   Expert systems

Based on the previous representation of knowledge (facts and rules), expert systems are a common way of programming. Table 6.2 (Turban and Aronson 1998), summarises the main differences between conventional programming systems and expert systems.

6.3.1   General structure of an expert system

In an expert system, knowledge is regrouped, often as scores of different pieces, in order to constitute a sort of 'program'. Expert 'rules' model behavioural or functional rules and 'facts' describe single values such as basic information or events. So, an expert system is an integration of a set of rules and a set of facts together with an inference engine (Figure 6.2). Notice the three parts of:

- the core of the expert system itself
- a module for knowledge acquisition
- a module interfacing the core with the user.

Note too that three categories of persons are usually involved in the design and the use of the expert system: the user, the subject matter expert and a knowledge engineer. **Knowledge engineering** is a process of codifying human knowledge. The role of the knowledge engineer is to hold discussions with the domain expert in order to extract her or his knowledge. This phase of knowledge acquisition from experts is crucial. The knowledge engineer's art is

Table 6.2 Comparison of conventional systems and expert systems (Turban and Aronson 1998)

Conventional systems	Expert systems
Information and its processing are usually combined in one sequential program	Knowledge base is clearly separated from the processing (inference) mechanism (that is, knowledge rules are separated from control)
Program does not make mistakes (programmers do)	Program may make mistakes
Do not (usually) explain why input data are needed or how conclusions are drawn	Explanation is part of most Expert Systems
Require all input data. May not function properly with missing data until planned for	Do not require all initial facts. Typically can arrive at reasonable conclusions with missing facts
Changes in the program are tedious	Changes in the rules are easy to make
The system operates only when it is completed	The system can operate with only a few rules (as for the prototype)
Execution is done on a step-by-step (algorithmic) basis	Execution is done by using heuristics and logic
Effective manipulation of large databases	Effective manipulation of large knowledge bases
Representation and use of data	Representation and use of knowledge
Efficiency is a major goal	Effectiveness is the major goal
Easily deal with quantitative data	Easily deal with qualitative data
Use numerical data representations	Use symbolic knowledge representations
Capture, magnify, and distribute access to numeric data or information	Capture, magnify and distribute access to judgement and knowledge

to pose the right questions in order to understand what knowledge the experts are using and what their ways of reasoning are and then to structure and encode that knowledge and logic.

When a new chunk of knowledge is going to be integrated into the knowledge base, some verifications are performed in order to check whether this chunk is consistent with the knowledge already included. Indeed, a new piece should not contradict the previous knowledge otherwise some corrections must be performed. Often, for the user's benefit, the expert system is completed with an explanation module in order to help users to understand the result of their acquisition of information from that knowledge base. The core of the expert system is the inference engine linked with the base of facts and the base of knowledge comprised of metarules and expert rules.

6.3.2 *Inference engine (Laurini and Thompson 1992)*

The role of the inference engine is to deduce, starting from input facts, some other facts, either intermediate or final output, using the encoded rules. There

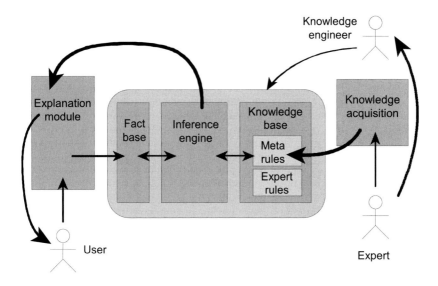

Figure 6.2 General structure of an expert system.

are several methods of reasoning that can be used. First we have deduction (*modus ponens*):

$$(P; P \rightarrow R) \Rightarrow R$$

meaning that if P is true, and that a rule $P \rightarrow R$ is also true, then we derive by deduction that R is true. Secondly, there is abduction (*modus tollens*), also called reasonable explanation:

$$(R; P \rightarrow R) \Rightarrow P$$

meaning that if R is true, and we have a rule stating that $P \rightarrow R$, then we obtain by abduction that P is true. Thirdly, we have reasoning by transitivity:

$$(P \rightarrow Q; \rightarrow QR) \Rightarrow (P \rightarrow R)$$

That is, if we have two different rules, the first implying Q and the second starting from Q, then by transitivity, a new rule can be produced, covering both P and R.

While expert systems may be based on these four types of logical reasoning, the more commonly used are sets of deductions, also called forward chaining, or sets of abductions called backward chaining, by means of transitivity. Forward chaining is the way to test the consequences of some starting context. It can be mainly used for 'what-if' reasoning. On the other hand, backward chaining is interesting for diagnosis, that is to discover the reasons generating the observed situation. For a toy example (Figure 6.3), the input facts are: A, B, C, D and E. Output facts are X and Y and the rules are:

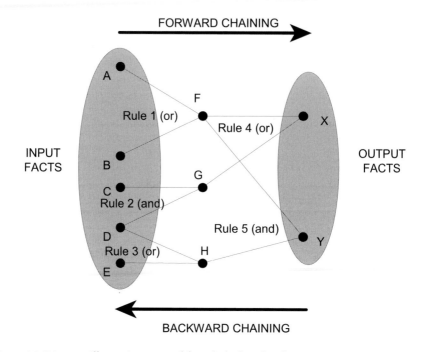

Figure 6.3 Diagram illustrating a set of facts linked with rules.
Laurini and Thompson, 1992.

Rule 1: (A or B) → F
Rule 2: (C and D) → G
Rule 3: (D or E) → H
Rule 4: (F or G) → X
Rule 5: (F and H) → Y

In this example, which also illustrates the conventional use of a graph structure for showing rules, we use Boolean conditions with OR and AND operations. However, in the majority of expert systems, only AND operators, known as Horn clauses, are used.

For this same example, in the case of forward chaining we have information about input facts, and we want to deduce information concerning the output facts. Suppose we know that B and D are true, A and C are wrong, and E unknown. By applying rule 1, we deduce that F is true, and by rule 4 that X is true. Moreover, rule 3 implies that H is also true, and since F is true, then Y is also true. Suppose now that A is true, B, D and E are false, and C unknown we can then deduce that F is true, G is unknown, and H is false. So X is true and Y is false.

Consider now that we impose values on the output facts, and that we are interested in knowing what the values of the input facts could be. In other words, we are looking for input values implying the observed output (diagnosis). As an example, suppose we look for a configuration so that X is true and Y is false. We can obtain (by abduction) immediately that:

F must be true
H must be false
G is indifferent

In this case, one of the solutions is:

A must be true
B and C must be indifferent
D and E must be false

In practical applications, sometimes a mixing of forward chaining and backward chaining is necessary. To elaborate a little on the rules, the previous example illustrates the case in which the rules are organised with a direct acyclic graph. Sometimes, though, loops between rules exist, causing the navigation of the expert system to be much more complex, and possibly producing some inconsistencies.

6.3.3   Example for landfill site selection

An interesting example is the problem of site selection, especially for disposing of refuse on land without creating nuisances or hazards to public health or safety (Rouhani and Kangari, 1987). For that, starting from Federal and State regulations, a database integrating the description of the soil and the subsoil, some algorithmic and rule-based models are run in order to output recommendations. See Figure 6.4 for details.

6.3.4   Conclusions about the use of expert systems in urban planning

Throughout the world, very few experiments have been done concerning expert systems for urban planning. First in the special issue of the Italian review (*Sistemi Urbani/Urban Systems*, XI, 4 April 1989) that I edited on expert systems for urban planning, I gave the description of some prototypes. Then, the book edited by Kim, Wiggins and Wright (1990) presented also an interesting state of the art. But practically all stayed at prototypical or academic levels, and, to my knowledge, none is daily used at operational level in urban planning because the main bottleneck is knowledge extraction, and especially rule extraction.

The second difficulty concerns spatial knowledge, or more exactly the integration of spatial topology and geometry into expert systems. Much has been done in this direction, but neither totally relevant nor effective approaches appear to exist at the moment. Nevertheless, the applications prototyped with expert systems ranged especially from site selection, traffic control to the resolution of environmental disputes. Some colleagues tried to begin research to find out rules for master plan design, but all have stopped

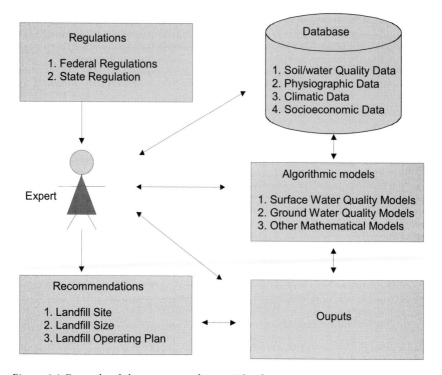

Figure 6.4 Example of the structure of a specialised expert systems.
Rouhani and Kangari 1987.

after some months, due to the impossibility, first to describe rules verbally, and then to formalise them as expert system rules.

According to Debenham (1998), knowledge based systems and expert systems are different. Expert systems perform in the manner of a particular trained expert, and they do not necessarily interact with corporate databases, and are closed in some manner. Whereas knowledge bases incorporate the knowledge relative to a domain, can be connected to databases, and finally can accept new applications, especially those based on reasoning techniques more complex than pure deductive reasoning.

6.4 Knowledge bases

Historically, after the failure of the expert system approach, some new paths were explored and pioneered under the name of knowledge engineering. The concept of knowledge was revisited, especially by defining several kinds of knowledge.

6.4.1 Different kinds of knowledge

There have been several papers which suggest how to identify knowledge in geographic and spatial systems; among them, let us mention McKeown *et al.* (1989) and Armstrong (1991). As a synthesis of those previous works, Crowther (1999) recently proposed the following classification:

- Primitive knowledge about the identification of primitives. A primitive is a readily identifiable point, line or area which cannot be subdivided into smaller named entities; this includes knowledge about an object's size and shape if relevant.
- Relationship knowledge of the spatial relationships between primitives in terms of their proximity, orientation and degree of overlap.
- Assembly knowledge, used to define collections of objects which form identifiable spatial decompositions; it includes knowledge of the spatial density of primitives; this knowledge can be regarded as knowledge needed for generalisation.
- Non-visual knowledge, which helps refine classifications including labelling of scene primitives and spatial relationships (spatial knowledge); it consists of temporal knowledge, algorithmic knowledge and heuristic knowledge.
- Consolidation knowledge used to resolve and evaluate conflicting information.
- Interpretation knowledge of how to combine the other knowledge types for understanding or reasoning.

These different kinds of knowledge come from different human or automatic experts. So it is necessary to validate and consolidate all the knowledge chunks.

6.4.2 Knowledge engineering

According to Debenham (1998), knowledge engineering is the design and the maintenance of knowledge-based systems. Design is a process that begins with an analysis of requirements and ends with a complete system specification. Given a change in circumstances, maintenance is the business of modifying the system specification to reflect those changes. A knowledge engineering methodology addresses the design and the maintenance of the knowledge as well as the information and the data. Figure 6.5 (from Debenham 1998) presents the main steps of this methodology starting from the application to the physical model, namely requirement model, conceptual model, functional model and internal model. This schema can be compared with Figure 2.2 presenting the ANSI-SPARC method to construct databases.

The various origins of knowledge are depicted Figure 6.6. Let us emphasise that not only facts and rules must be stored but also, processes, constraints, heuristics, decision rules, procedures for problem solving, etc. In some

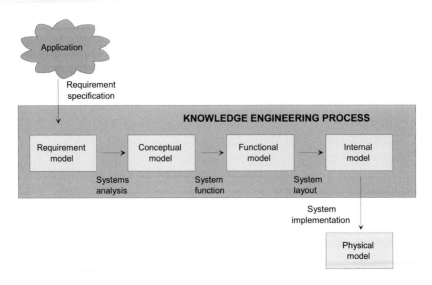

Figure 6.5 The knowledge engineering process.
According to Debenham 1998.

knowledge bases also typical situations and best practices are stored in order
to create an organisational memory (O'Leary 1998).

6.4.3   Metarules

Besides rules, an expert system can possess metarules. A metarule is a rule
concerning knowledge, for instance, a procedure for selecting other rules.
Often, strategies for solving problems are included in metarules. For example,
IF such a condition occurs, THEN APPLY Rule 234 and Rule 456. When
designing an expert system it is important to delineate metarules because their
usage can accelerate logical inferences.

Figure 6.7 depicts several organisations for rule bases. Figure 6.7a is given the
conventional sequential rule base whereas 6.7b illustrates the sequential case
with certainty factors (CF); for instance CF = 80 %. A first amelioration can be
found by sorting all rules according to the frequency of use (Figure 6.7c). So, the
first checked rules are the more commonly invoked. But the paramount
amelioration is made by using metarules as illustrated in Figure 6.7d.

6.4.4   Spatial metarules

Generally speaking, using the areal spatial units as examples, spatial rules can
be applied to several zones and a zone can have several rules. Figure 6.8
depicts an example of relationships between zones and spatial rules: zone 1 has

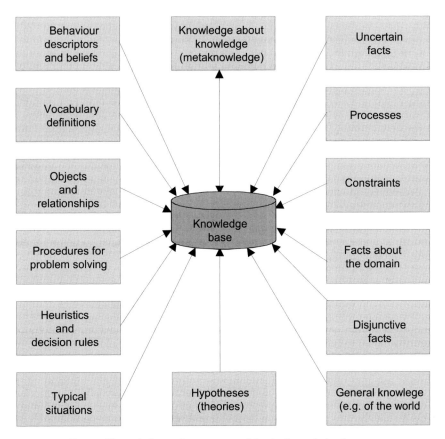

Figure 6.6 Type of knowledge to be represented in the knowledge base. According to Turban and Aronson 1998.

rule B; zone 2 has rules A and C; zone 3 has rules C and D, and zone 4 has rule B. And so one task is, given an (x, y) point, to ascertain the applicable spatial rules. This task first requires a point-in-polygon query to be solved; and then metarules can be defined:

IF (x, y) belongs to Zone 1 THEN rule B is applicable.
IF (x, y) belongs to Zone 2 THEN rules A or C are applicable.
IF (x, y) belongs to Zone 3 THEN rules C or D are applicable.
IF (x, y) belongs to Zone 4 THEN rule B is applicable.

6.4.5 Knowledge consolidation and validation

In a lot of cases, knowledge chunks can come from totally different sources. Facing such an issue, some consolidation or validation procedures must be

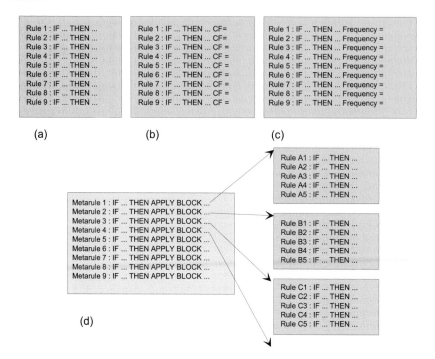

Figure 6.7 Several organisations of a rule base. (a) The sequential organisation. (b) The sequential organisation with certainty factors. (c) The sequential organisation in which rules are sorted by frequency. (d) The metarule organisation.

run. Table 6.3 from Turban and Aronson (1998) shows some elements of this very important step in knowledge base design. A special case of knowledge is represented by argumentation; see Section 6.6 and Chapter 9 for details.

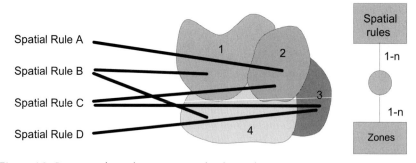

Figure 6.8 Correspondence between spatial rules and zones.

Table 6.3. Measures of validation of a knowledge base (Turban and Aronson 1998 p. 516)

Measure (criteria)	Definition
Accuracy	How well the system reflects reality, how correct the knowledge is in the knowledge base
Adaptability	Possibilities for future development changes
Adequacy (completeness)	Portion of the necessary knowledge that is included in the knowledge base
Appeal	How well the knowledge base matches intuition and stimulates thought and practicability
Breadth	How well the domain is covered
Depth	Degree of the detailed knowledge
Face validity	Credibility of knowledge
Generality	Capability of a knowledge base to be asked with a broad range of similar problems
Precision	Capability of the system to replicate particular system parameters, consistency of advice, coverage of variables in knowledge base
Realism	Accounting for relevant variables and relations, similarity to reality
Reliability	Fraction of the expert system predictions that are empirically correct
Robustness	Sensitivity of conclusions to model structure
Sensitivity	Impact of changes in the knowledge base on quality of outputs
Technical and operational validity	Quality of the assumed assumptions, context, constraints and conditions, and their impact on other measures
Turing test	Ability of a human evaluator to identify whether a given conclusion is made by an expert system or by a human expert
Usefulness	How adequate the knowledge is (in terms of parameters and relationships) for solving correctly
Validity	Knowledge base's capability of producing empirically correct predictions

6.5   Ontologies (Gruber 1991, 1992)

According to Gruber, an ontology is a specification of a conceptualisation. Ontologies are often equated with taxonomic hierarchies of classes (for instance) but class definitions, and the subsumption relation, but ontologies need not be limited to these forms. Ontologies are also not limited to conservative definitions, that is, definitions in the traditional logic sense that only introduce terminology and do not add any knowledge about the world. To specify a conceptualisation one needs to state axioms that constrain the possible interpretations for the defined terms.

In the context of knowledge sharing, the term ontology is used to mean a specification of a conceptualisation. That is, an ontology is a description (like a formal specification of a program) of the concepts and relationships that can exist for an agent or a community of agents. This definition is consistent with the usage of ontology as set-of-concept-definitions, but more general. It is certainly a different sense of the word than its use in philosophy.

For pragmatic reasons, an ontology can be seen as a set of definitions of formal vocabulary. Although this is not the only way to specify a conceptualisation, it has some nice properties for knowledge sharing (e.g., semantics independent of reader and context). Practically, an **ontological commitment** is an agreement to use a vocabulary (i.e., ask questions and make assertions) in a way that is consistent (but not complete) with respect to the theory specified by an ontology. One can build agents that commit to ontologies. So, one can design ontologies in order to share knowledge with and among these agents.

Pragmatically, a common ontology defines the vocabulary with which queries and assertions are exchanged among agents. Ontological commitments are agreements to use the shared vocabulary in a coherent and consistent manner. The agents sharing a vocabulary need not share a knowledge base; each knows things the other does not, and an agent that commits to an ontology is not required to answer all queries that can be formulated in the shared vocabulary. In short, a commitment to a common ontology is a guarantee of consistency, but not completeness, with respect to queries and assertions using the vocabulary defined in the ontology.

Figure 6.9 gives an example of an ontology for underground water resources (Hadzilacos *et al.* 1999). Figure 6.10 illustrates the relationships between several types of ontologies; an ontology for a specific application must originate from two different ontologies, domain- and task-ontologies, all issuing from a very general one (top-level ontology).

An example is featured Figure 6.11 in which two applications need different definitions of a parcel. By means of ontologies, the concepts and the vocabulary are well defined. According to Chandrasekaran and Josephson (1999), there is a general agreement between ontologies on many issues:

- there are objects in the word
- objects have properties or attributes that can take values
- objects can exist in various relations with each other
- properties and relations can change over time
- there are events that occur at different times
- there are processes in which objects can be in different states
- the world and its objects can be in different states
- events can cause other events or states as effects
- objects can have parts.

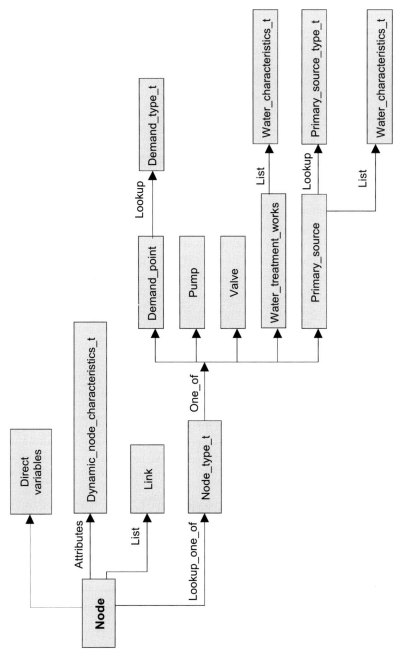

Figure 6.9 An excerpt from an ontology approximation on underground water resources. From Hadzilacos *et al.* 2000.

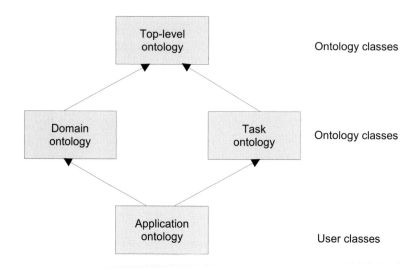

Figure 6.10 Ontologies organisations.
From Torres-Fonseca and Egenhofer (2000) after Guarino *et al.* (1994).

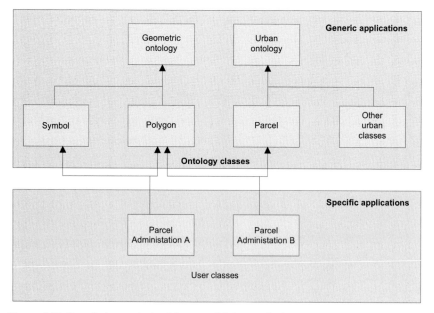

Figure 6.11 Parcel classes derived from multiple ontologies.
After Torres-Fonseca and Egenhofer (1999).

For an example of ontologies in cartography, please refer to Andrienko and Andrienko (1999).

In Chapter 10, the problem of GIS interoperability is addressed, and several hopes rest on ontologies as a key issue for real-time data exchanging taking semantic differences into account. Now, let us pass to the modelling of a very important type of knowledge, i.e. debate and argumentation modelling.

6.6 Debate and argumentation modelling

During public participation stages (Chapter 9), it is important first that the citizens can understand what is being proposed to them, and secondly that the planners understand what the citizens are proposing. So debate and argumentation modelling is an important issue in any participative planning process. The goal of this section will be to provide the readers basics for this kind of modelling. However, this issue is also important for all actors involved in the co-operative planning process (see Chapter 8).

According to Tweed (1998), the potential benefits of recording the argumentation and debate surrounding different types of decision making have been recognised in all activities which bear some relation to urban planning and design. Recorded argumentation can explain why particular decisions were taken and so has the capacity to justify recent decisions as well as inform future decision making and the formulation of new policy. Decision making in urban planning is inherently complex because it embraces so many different interests. It would appear, at least on the surface, to be ripe for the development of argumentation systems, since increasing distrust of technical rationality among the lay public, coupled with the demand for greater transparency, openness and accountability of decision-making processes. There is a much greater need to explain and justify decisions than before. Argumentation would appear to have something to offer in assisting public participation in urban planning and design.

In the UK, recording argumentation, particularly in relation to design services, will in part satisfy the quality assurance requirements of BS 5750 Part 1 (Tweed 1998), which addresses the need for design consultancies to manage their provision of services to their clients. To achieve certification design consultancies have to keep records of all design decisions and changes. Different methods have been proposed to describe argumentations. Let us detail them.

6.6.1 From Toulmin to Gottsegen

Toulmin's model (1958) of argumentation is simple in structure. It has three components: claims, data and warrants (Figure 6.12). Data are assertions of any type. They may be given from some external source, or they may be the result of arguments themselves. Data support claims which are conclusions of an argument. Warrants justify the inference of claims from data. The data and warrants can also have backing in the form of additional data. Each of these

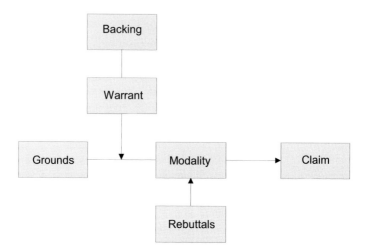

Figure 6.12 Toulmin's structure for representing argumentation.

roles can be the target of an attack or counterargument (Gottsegen 1998). An attack is an assertion that weakens the function of its target until it is effectively answered. Of course, counterarguments may be data or claims themselves and often require backing. For the sake of clarity, this section refers to data-warrant-claim units as 'subarguments'. Such subarguments are chained together to form larger threads of argument. One uses the term 'argument' to refer to the macro structure of argumentation, i.e., the entire debate (this differs from the definition

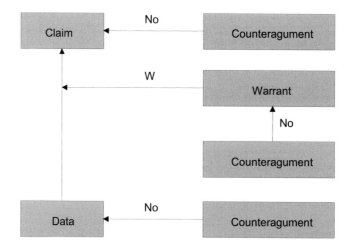

Figure 6.13 Gottsegen's model with counterarguments (Gottsegen 1998).

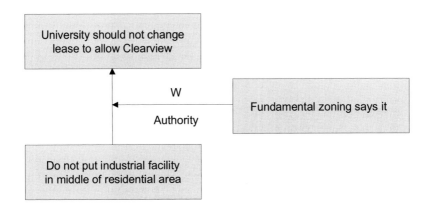

Figure 6.14 Example of subargument.

of argument in formal logic). Figure 6.13 shows how the argumentation roles can be depicted graphically. The process of argumentation analysis assigns these roles to assertions thereby indicating their function in the argument relative to other assertions. Figure 6.14 gives an example of a subargument, whereas a more complete example is depicted Figure 6.15 which represents the main arguments and counterargument related to a debate concerning the 'Clearview Debate', i.e; the construction of a slant drilling facility to retrieve off-shore oil linked to an on-shore site owned by the university of California at Santa Barbara (Gottsegen 1998). An excerpt of the complete debate model is given Figure 6.16.

6.6.2   From IBIS to Tweed (1998)

IBIS[2] (Issue-Based Information Systems), developed by Kunz and Rittel (1970), centres on the explication of issues. Issues are at the core of debate represented in IBIS and related issues are grouped under a 'topic' in an effort to keep things manageable. Figure 6.17a illustrates the relationships between ideas and some justifications, starting from a possible question. Based on this diagram, Figure 6.17b give the basics of IBIS in which justifications are replaced by argument. So, all conversations or issues in IBIS consist of:

- 'Question' states a question
- 'Idea' proposes a possible resolution for the question; and
- 'Argument' states an opinion or judgement that either supports or objects to one or more ideas.

---

2   For more information about IBIS, please refer to http://www.gdss.com/IBIS.htm, *'The IBIS Manual, A Short Course in IBIS Methodology'*.

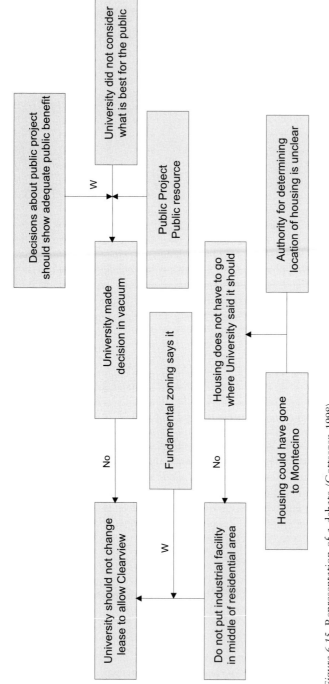

Figure 6.15 Representation of a debate (Gottsegen 1998).

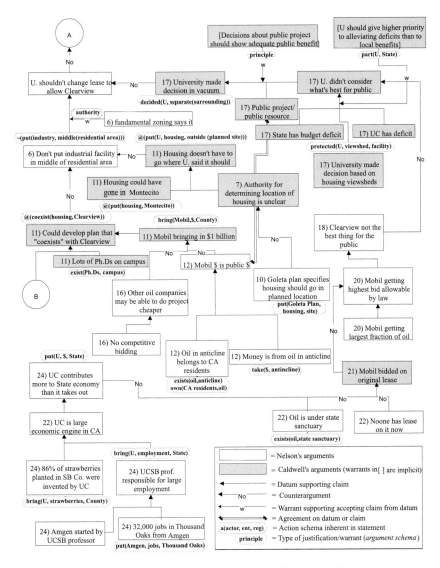

Figure 6.16 Analysis of portion of Clearview Debate (Gottsegen 1998).

Issues are usually expressed in question form, for instance 'is subsidised public transport the most cost-effective way to reduce urban traffic'? Responses to such questions generate 'positions' which are linked to the issue by *responds-to* arcs. Positions are established, justified or attacked by 'arguments'. IBIS in its original formulation, therefore, provides a set of typed nodes and links which may be used to construct networks of

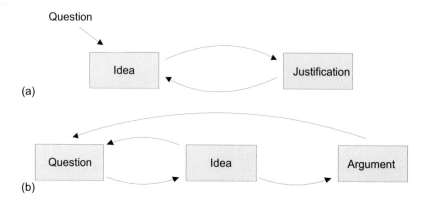

(a)

(b)

Figure 6.17 Basics of IBIS. (a) Feedbacks between ideas and justification. (b) Links between question, idea and arguments.

argumentation. Figure 6.18 shows the principal node and link types offered by IBIS.

The original formulation of IBIS offers a rigorous framework which defines logical properties of the elements as well as a method of operation for the system as a series of manual procedures. Since then others have adopted the method, whole or in part, and implemented computer programs to apply it to a variety of domains, notably design. In most cases the method of operation has been discarded, and what remains is the set of elements alone. The gIBIS system (Conklin and Begeman 1988) is well documented as an example of IBIS implemented as a system to capture

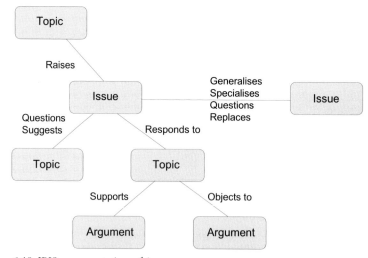

Figure 6.18 IBIS representation of issues.
According to Tweed (1998).

170

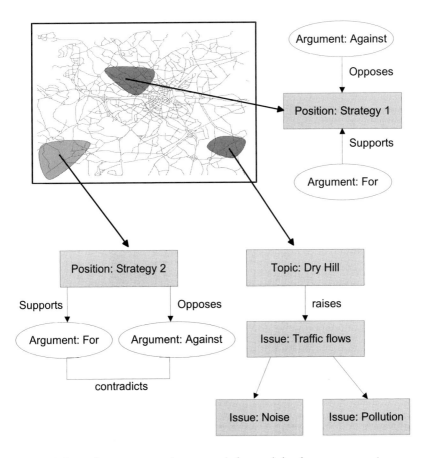

Figure 6.19 Part of an argumentation network for road development strategies.

design rationale in different domains. The main features of the system are a graphical browser, structured index, inspection window and control panel. Networks of arguments are displayed as graphical elements depicting nodes and links and associated textual information, such as node names, types, owners, etc. Each node may also have an associated content, such as a text file.

Based on gIBIS, Tweed (1998) has combined this model with hypermaps by using *CrossDoc* the principle of which is given in Figure 6.19. Moreover, Figure 6.20 shows part of an argumentation network constructed to illustrate the structuring of a debate between multiple stakeholders about increases in noise levels which may follow further development of a motorsport racetrack.

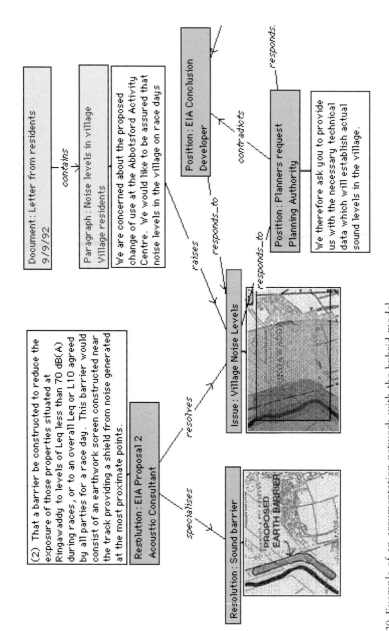

Figure 6.20 Example of an argumentation network with the hybrid model.
Reprinted from *Computers, Environment and Urban Systems* 22 4, C. Tweed 'Supporting Argumentation Practices in Urban Planning and Design' 351–63 © 1998, with permission from Elsevier Science.

6.6.3   Argumaps (Rinner 1999)

Recently, Rinner (1999) has proposed a new concept named *Argumaps* (Argumentation Maps) which is based on argumentation and hypermaps (Chapter 5). Participants of spatial planning discussions refer verbally to geographic objects. For example, an argument supporting a new industrial zone refers to the appropriate map location. An argument against the construction of a new highway refers to the distance between the planned route and a housing area. Rinner has proposed an explicit linkage between online maps and discussion contributions for being used in a World-Wide Web-based support system for collaborative spatial decision-making.

The representation and storage of geo-referenced arguments in Argumaps would advance the level of integration and utility of electronic discussion forums and digital plans. Argumaps will provide graphical tools for visualising geo-referenced contributions and for interactively following links between arguments and map objects. Thus, users involved in public planning debates have a navigable cartographic 'index' to a discussion that enables them to explore spatial structures in the arguments. In addition, functions for querying and analysing geo-argumentative relations and for facilitating the submission of constructive contributions will be available. The quality of planning discussions will benefit from an improved retrieval and use of available documents.

In a prototype designed and implemented by Rinner, the Descartes system (http://allanon.gmd.de/and/java/iris/) is used for producing intelligent, interactive maps for the visual exploration of geo-referenced statistical data (Figure 6.21). The client/server system (see Chapter 10) rests upon a knowledge-base with rules for cartographic design and builds maps according to the characteristics of the thematic data at hand. The user can change the map display interactively, e.g. by manipulating the colour scale for data values or by modifying the default classification. Visual exploration of data sets means a lot of 'playing' with the map display, to find peculiarities in spatial structures.

Aggregated data about a plan-related discussion can be visualised by Descartes like any geo-referenced data set. Figure 6.22 shows the distribution of the difference in number of pro arguments minus contra arguments per geographic object, assuming that the map features were planning areas, linked to contributions of discussion participants. Two modes of locating the arguments are shown, by pin-pointing (Figure 6.22a) or with flags (Figure 6.22b). In conjunction with the colour scale, the map enables the viewer to capture immediately what regions have been more (red/dark grey) or less (green/light grey) disputed than the white reference area, and which area has been most contradicted.

In Descartes, it is possible to save an interactive map visualisation as a stand-alone Java applet. This is interesting if an exploration session precedes participation when preparing a selective contribution. Such an interactive map

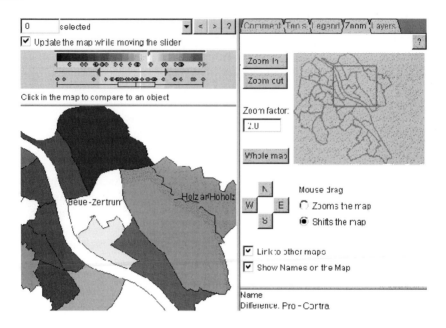

Figure 6.21 Argumap exploration with Descartes (Rinner 1999).
Published with permission.

can easily be included in an HTML message sent to the discussion forum, in
order to underline a statement of the author. This system can easily be
extended to store vocal arguments.

6.7   Metadata and metainformation

According to Devlin (1997), 'metadata are data that describe the meaning and
structure of business data, as well as how it is created, accessed, and used'.
They are a standardised way to describe features (classes, collections, objects,
attributes, etc.) as stored in the database. Moreover, metadata must describe
the data sets in sufficient detail that the user can evaluate the fitness of use of
the data within his applications. Similarly for business or planning data,
metadata can be classified along some basic criteria. Devlin distinguishes three
types of metadata:

1   **Build-time metadata** which describes the metadata created and used in the
    process of application and database design and construction. Even when
    an application is upgraded, only a small proportion of the metadata is
    likely to change. Metadata describing planning meaning may be stable
    over a period of years, depending on the domain in which the organisation
    operates.

174

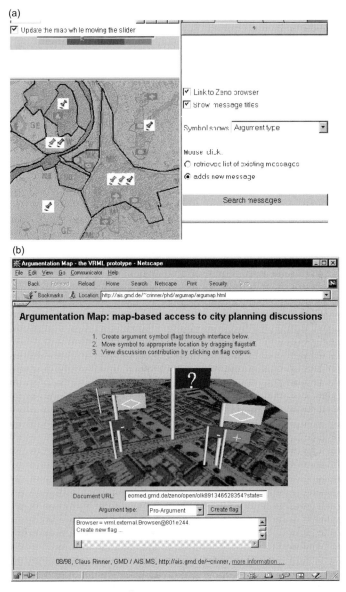

Figure 6.22 Navigation session with Descartes (Rinner 1999). (a) By pin-pointing arguments. (b) With flags.
Published with permission.

2 **Control metadata** refers to metadata that are actively used to control the operations related to the database or the datawarehouse (see Section 10.6). Two kinds can be distinguished; timeliness metadata and utilisation

175

Table 6.4. The Metadata environment (O'Brien 1999)

	Catalogue	Where Within datasets	Historical records	Textual reports
Applications	Discover Evaluate Access Exploit			

metadata. Timeliness metadata describes the actual information about the timeliness of planning data, such as timestamps. Utilisation metadata is most closely associated with security and authorisation.

3   **Usage metadata** is the most important type of metadata for the informational environment, because it is structured to support end users' use and understanding of planning data. Four levels exist:

(a) meaning of the planning data
(b) ownership and stewardship
(c) data structure
(d) application aspects.

According to O'Brien (1999), Table 6.4 describes the environment of metadata, detailing the kind and location of description *vis-à-vis* concerned applications. This table emphasises that there exist multiple ways of storing metadata. This table was made in the context of the Canadian Geospatial Data Infrastructure[3] for the description of information as a resource for which three steps can be distinguished:

• resource discovery
• resource evaluation
• resource access.

Finally, any user must access a metadatabase, i.e. a base storing metadata, in the same way that he is dealing with common databases (Figure 6.23).

So, usage metadata must be organised very clearly. For this purpose, for each data item, one must collect the definition, the documentation, the way it was created as so on. Several possibilities exist. The first one is to create a dictionary of all data items, a data dictionary or classification register when it is a sort of linear list. The second possibility is to confer metadata a structure similar to a conventional database because several metadata entities or objects can be considered. A conceptual model or an object model is shown in Figure 6.24 from Barquin and Edelstein (1997).

---

3   http://ceonet.cgdi.gc.ca

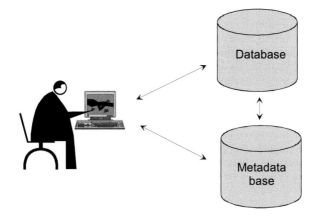

Figure 6.23 Relationships between the user, the database and the metadatabase.

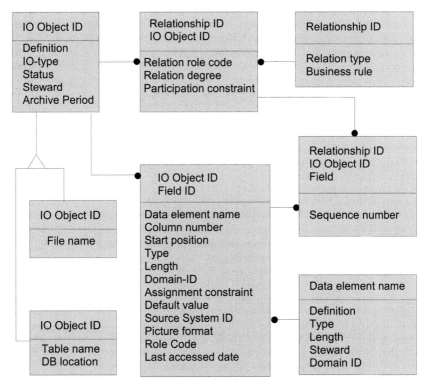

Figure 6.24 Metadata model describing the contents of the datawarehouse.
After Barquin and Edelstein (1997) with modifications.

The major difficulty is not to design the conceptual model, but to capture metadata. In fact metadata is very difficult to capture, and a sort of bottleneck of metadata now commonly exists. Due to the vocabulary in use, some people are considering the use of thesauri or ontologies in order to structure the vocabulary, so that the real meaning of the objects are included into the database.

6.8 Spatial metainformation

According to Balfanz and Göbel (1999), here is a quick story of spatial metadata. On 11 April 1994, President Clinton signed the 'Executive Order 12906, Coordinating Geographic Data Acquisition and Access: The National Spatial Data Infrastructure'[4] in order to co-ordinate data acquisition and access to geospatial data in the USA. This executive order instructs federal agencies to document new geospatial data using the metadata standard under development by the FGDC (Federal Geographic Data Committee),[5] and to make these metadata available to the public through the National Geospatial Clearinghouse. In June 1994, the first version of the 'Content Standards for Digital Geospatial Metadata' (FGDC/CSDGM) was approved, in June 1998 version 2. The CSDGM Version 1 has also been the initial draft for the development of ISO (International Organization for Standardization)[6] metadata standard 15046-15, carried out by the Technical Committee 211.

Among others also the OGC (Open GIS Consortium)[7] contributes the results of its endeavours to ISO/TC 211. Furthermore there is a so-called class A liaison between ISO and OGC, which aims to harmonise the plans of OGC and TC211 in the field of geoprocessing standards and guarantees the mutual use of experience/results. Neither on a European level nor on a national level is there a comparable regulating law or an executive order as in the USA. A lot of different metadata formats and information systems are emerging all over Europe. But there is also a remarkable trend to harmonise the efforts. Some current initiatives shall be mentioned.

MEGRIN[8] (Multipurpose European Ground Related Information Network), an organisation representing and owned by 19 European NMAs (National Mapping Agencies), established the Geographic Data Description Directory in order to help users of geographic information by simplifying access to digital map data of NMAs. MEGRIN itself is an initiative of CERCO (Comité Européen des Responsables de la Cartographie Officielle),

---

4   Executive Order 12906: http://www.fgdc.gov/publications/documents/geninfo/execord.html.
5   http://www.fgdc.gov.
6   http://www.statkart.no/isotc211.
7   http://www.opengis.org
8   http://www.megrin.org/gddd/gddd.htm

the forum for heads of European NMAs. MEGRIN's role was to develop a Geographical Data Description Directory, named GDDD.

The GDDD is based upon the European standard CEN/ENV 12657 – Data Description – Metadata,[9] which was approved in October 1998 and is also part of the ESMI project (European Spatial Metadata Infrastructure). ESMI is partly funded by the European Commission as a project of the INFO2000 program. The origin of the initiative is settled in the goal of several public and private organisations in Europe to establish a framework in the form of a universal metadata service for geographic information. Existing metadata systems should be linked and future systems should be easily integrated into this universal system too. Let us now present more carefully the FGDC and the ISO standards.

6.8.1   The Federal Geographic Data Committee (FGDC) metadata standard

Many organisations that manage geographic databases find it useful to capture and update 'metadata' about the geographic data. Metadata includes information about the content, format, quality, availability, and other characteristics of a GIS database. Metadata can help answer such questions from database administrators and users as 'what data is available?', 'what are its characteristics and quality?', 'how do I access it?', 'will it meet my needs?' Metadata are designed to facilitate such database activities as

- on-going update
- browsing and discovery
- distribution
- proper use in applications.

A metadatabase can be thought of as a robust dictionary to a geographic database which makes the GIS data easier to manage and use.

The Federal Geographic Data Committee (FGDC) has developed a standard for storing metadata called the Content Standard for Geospatial Metadata. This standard lays out major categories of metadata and specific data items that may be stored in a metadatabase and it provides a model for user organisations to adopt. Figure 6.25 provides an overview of the types of information that is stored in a GIS metadatabase.

In order to facilitate information retrieval, it was necessary to set up different user profiles according to some expertise level. This approach has led Foresman

---

9   http://www.cenorm.be

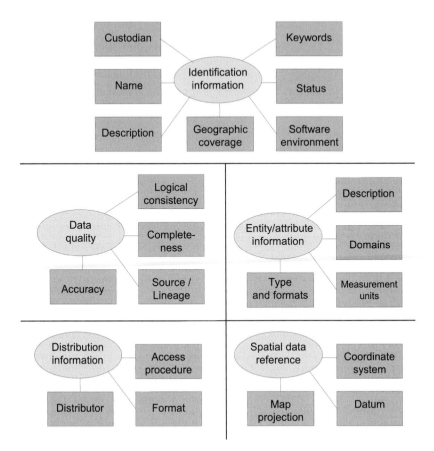

Figure 6.25 Overview of the FGDC Metadata Content Standard.

et al. (1996) to propose three levels of metadata associated to three levels of users (Figure 6.26):

1 simple users who do not need very detailed information regarding data
2 decision makers including scientists and managers who have a very broad knowledge of the domain, and
3 experts who master the specifications and the structure of information.

So, more important is expertise, more complex and detailed are the metadata.

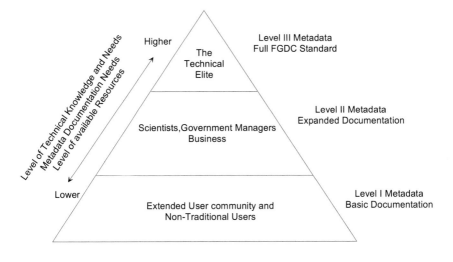

Figure 6.26 Levels of metadata.
Introduced by Foresman *et al.* (1996).

6.8.2 The ISO standards for metadata

The International Organization for Standardization (ISO) proposes a specific standard for quality and metadata including the following elements[10] (part 15 on metadata):

- Identification information: Language of dataset code, Dataset character code set, Abstract, Purpose, Supplemental information, Dataset environment, Dataset credit, Identification citation, Initiative identification information, Image identification information, Status, Dataset extent, Resolution level code, Category, Primary point of contact, Browse graphic, Dataset association, Dataset constraints
- Data quality information: Data quality
- Lineage information: Usage, Lineage
- Spatial data representation information: Spatial representation type code,

---

10 ISO 15046 is a multi-part International Standard under the general title Geographic information consisting of the following parts: Part 1: Reference model, Part 2: Overview, Part 3: Conceptual schema language, Part 4: Terminology, Part 5: Conformance and testing, Part 6: Profiles, Part 7: Spatial schema, Part 8: Temporal schema, Part 9: Rules for application schema, Part 10: Feature cataloguing methodology Part 11: Spatial referencing by coordinates, Part 12: Spatial referencing by geographic identifiers, Part 13: Quality principles Part 14: Quality evaluation procedures, Part 15: Metadata, Part 16: Positioning services, Part 17: Portrayal, Part 18: Encoding, Part 19: Services.

Vector spatial representation information, Raster spatial representation information, Image spatial representation information

- Reference system information: Temporal reference system information, Spatial reference information
- Feature catalogue information: External feature catalogue, Internal feature catalogue
- Distribution information: Distributor information, Distribution format information
- Metadata reference information: Metadata file identifier, Metadata parent identifier, Dataset application schema name, Metadata date information, Metadata contact, Metadata standard information, Metadata extension information, Metadata ancillary information, Metadata constraints, Metadata security constraints

6.8.3   Example for the lagoon of Venice, Italy (Hedorfer and Bianchin 1998, Bianchin 1999)

This example is taken from (Hedorfer and Bianchin 1998, Bianchin 1999) for the lagoon of Venice, Italy, for which it was necessary to create not only a database including all types of information, but also a metadata repository (Figure 6.27).

In this example, it was necessary to define a metadata set to carry out the inquiry into the data available at different public and private authorities. As a result, a frame of parameters was proposed and implemented to carry on the analysis of data found. Such a frame gave form to a standard questionnaire divided into nine groups of questions related to the following features.

- the nature of data: image or text, that means whether data is spatially referenced and structured or not
- geographical area concerned and the scale of reasoning
- elements defining data: classification (nominal, discrete, continuous); finality (descriptive or management), spatial settlement (point, line, area); localisation accuracy, measurement accuracy
- temporal attribute
- quality assessment
- sources of data and methodology of acquisition
- kind of support and format if it is a digital one
- availability of data
- any other information of data.

Secondly, the metadata gathered following the frame mentioned above, has been organised in a relational database. The objective of that second step is to supply a tool able to support the search of data and in particular answer such question as: what data is available, who owns it, what kind of data am I faced with, how can I retrieve it, how reliable is it, does it fit my purpose, which exchange formats are available, who can I contact for further details? At this stage it was realised that it was a waste to build a database just for their own use and that it would be more useful to make that database available to a wider number of users through the Internet.

Designing and implementing a **web site for the lagoon of Venice** has been the second step of the project. The web site is designed to include many services able to give detailed information on the lagoon of Venice: administrative information, scientific, planning, etc; real-time data like tidal events concerning the lagoon, etc. A prototype was achieved with two services; the navigation and the query into the list of laws concerning the lagoon and into the metadata database.

The laws service

This service supplies all laws, national, regional or specific to the lagoon. Each law is a text with its title and a short abstract describing the content at the top. Some key words, expressed by a numeric code, are associated with each law in order to classify the laws according to some typologies. Other components of the middleware (Chapter 10) allow searching of any part of text and/or cross it with a key word and to create the output on the web page. The design of the web interface provides two points of access which can be activated simultaneously:

- by law title (regional or national), which can be selected through a top-down menu
- by key word.

The metadata database and its query

On the basis of the metadata frame described above and considering the metadata elements collected – note that some parts of the questionnaire had very few answers – the relational database has been organised in six tables:

- Table 1 containing general data of the project or database: titles, organisation, responsibilities, references to laws, contacts, author of the questionnaire
- Table 2 containing the characteristics common to the set of data such as scale, nature of data, origin of data, etc.
- Table 3 containing the themes (later called variables) with all the characteristics affecting them

- Table 4 containing information about physical supports and formats
- Table 5 containing different dates of capture, production and publication of data
- Table 6 containing all the information for the acquisition of the data and bibliography.

The next problem was how to structure the gathered metadata to make its search and visualisation on the Web as smooth as possible. To that purpose, reference was made to the five functions quoted in the literature: Browsing, Search, Retrieval, Transfer and Evaluation, on which basis were planned the system's architecture and the input and display visualisation of metadata.

In the lagoon case, those same functions contributed to the definition of five sections in which the gathered metadata was grouped, especially in view of Web queries. Actually, the metadata database is viewed from the web user following the five sections corresponding to the approximate sequence of the search process. They are defined as follows:

- General data, containing general information about the data context, i.e. the project or database in which they are included. This information concerns:
  - Source, Author, Agency (who owns the data)
  - Title and General Project (the name of the project or database where data is to be found)
  - Information Representation (how the data is presented: as number, image or text)
  - Georeferentiation (whether or not the data is georeferenced)
  - Geographic Extent (the spatial area covered by data)
  - Scale (level of detail or level of reasoning).

Now, they deal with the problem of supplying the user with information on the context of data use, in order to permit evaluation of the semantic aspects of the data, or penetration in the 'significance' of the data. This section is mostly related to the browsing and search functions:

- Variables: this is the list of the data (variables) included in a project. For each datum information is supplied on:
  - Typology (nominal, discrete, continuous)
  - Use (descriptive, managerial)
  - Geographic Entity (point, line, area)
  - Key Word (defines the thematic area to which the data make reference).

The Key Word makes reference to the variable rather than to the project. This allows reaching the desired data by cutting across the projects and bypassing the need to use exactly the same term for the Variables in each project.

- Availability and supports: this section supplies information about data retrieval by specifying if the data is:
  - purchasable
  - available upon convention
  - not available
  - support available (analog or digital)
  - recording format
  - available formats for interchange.
- Data quality and data timing: this section contains information about the data quality, expressed through:
  - population and survey samples (whether the data originated from digital photograph, analog photograph, numeric map, analog map, grid of points, census, administrative data or other)
  - acquisition modes (how the data was captured and measured)
  - types of control carried out (this concerns quality control both for the procedures of data construction and for the equipment used)
  - quality responsible (within the structure or agency which built the data)
  - measurement timing (whether systematic: hourly, daily, weekly, monthly, etc., or occasional)
  - beginning and ending dates (when dealing with data taken occasionally)
  - dates: of data, of publication, of next Collect, digitizing (when dealing with data taken occasionally).

The metadata described here reflects the specific situation of the data collected on the lagoon. In many cases these are included in planning documents and represent analyses finalised at planning. The highly variable ways of their collection obviously influence the data quality and significance.

- Bibliography and annotations: this section supplies information on possible bibliographic references (in the case of published data) and any other note which may contribute to a better definition of the specific data. This is actually a space reserved for any information deemed useful as a complement to the metadata: not only texts, but also schemes, graphs, maps or other images.

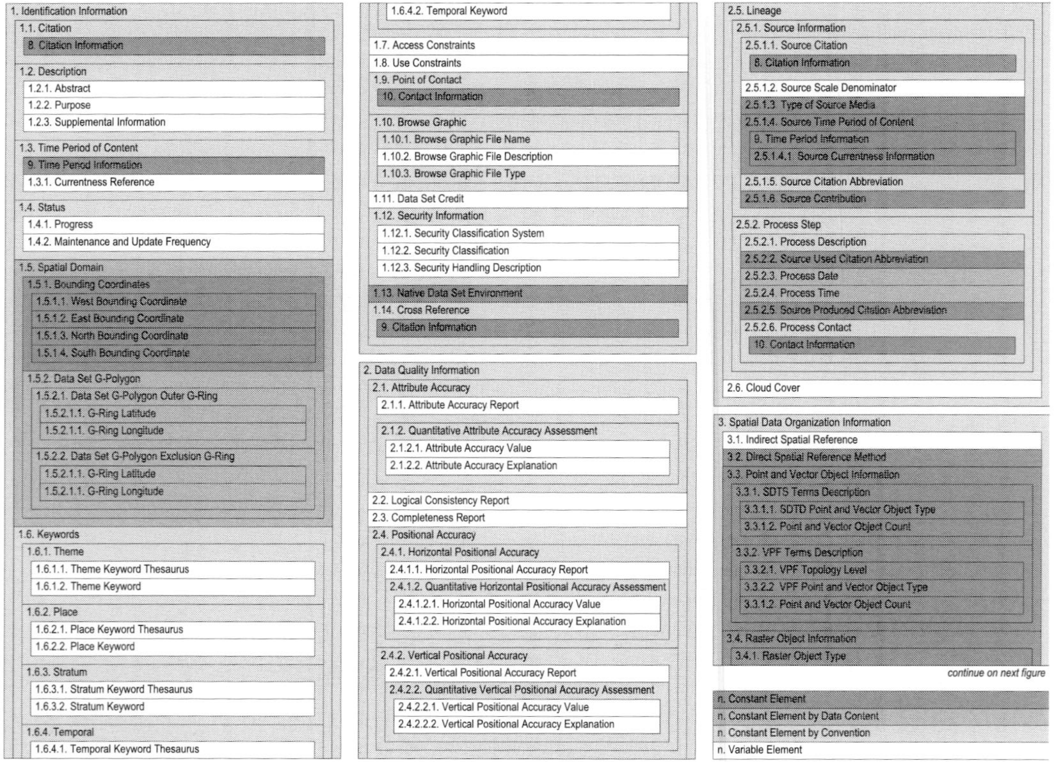

1. Identification Information

1.1. Citation
- 8. Citation Information

1.2. Description
- 1.2.1. Abstract
- 1.2.2. Purpose
- 1.2.3. Supplemental Information

1.3. Time Period of Content
- 9. Time Period Information
- 1.3.1. Currentness Reference

1.4. Status
- 1.4.1. Progress
- 1.4.2. Maintenance and Update Frequency

1.5. Spatial Domain
- 1.5.1. Bounding Coordinates
  - 1.5.1.1. West Bounding Coordinate
  - 1.5.1.2. East Bounding Coordinate
  - 1.5.1.3. North Bounding Coordinate
  - 1.5.1.4. South Bounding Coordinate
- 1.5.2. Data Set G-Polygon
  - 1.5.2.1. Data Set G-Polygon Outer G-Ring
    - 1.5.2.1.1. G-Ring Latitude
    - 1.5.2.1.1. G-Ring Longitude
  - 1.5.2.2. Data Set G-Polygon Exclusion G-Ring
    - 1.5.2.1.1. G-Ring Latitude
    - 1.5.2.1.1. G-Ring Longitude

1.6. Keywords
- 1.6.1. Theme
  - 1.6.1.1. Theme Keyword Thesaurus
  - 1.6.1.2. Theme Keyword
- 1.6.2. Place
  - 1.6.2.1. Place Keyword Thesaurus
  - 1.6.2.2. Place Keyword
- 1.6.3. Stratum
  - 1.6.3.1. Stratum Keyword Thesaurus
  - 1.6.3.2. Stratum Keyword
- 1.6.4. Temporal
  - 1.6.4.1. Temporal Keyword Thesaurus
  - 1.6.4.2. Temporal Keyword

1.7. Access Constraints

1.8. Use Constraints

1.9. Point of Contact
- 10. Contact Information

1.10. Browse Graphic
- 1.10.1. Browse Graphic File Name
- 1.10.2. Browse Graphic File Description
- 1.10.3. Browse Graphic File Type

1.11. Data Set Credit

1.12. Security Information
- 1.12.1. Security Classification System
- 1.12.2. Security Classification
- 1.12.3. Security Handling Description

1.13. Native Data Set Environment

1.14. Cross Reference
- 9. Citation Information

2. Data Quality Information

2.1. Attribute Accuracy
- 2.1.1. Attribute Accuracy Report
- 2.1.2. Quantitative Attribute Accuracy Assessment
  - 2.1.2.1. Attribute Accuracy Value
  - 2.1.2.2. Attribute Accuracy Explanation

2.2. Logical Consistency Report

2.3. Completeness Report

2.4. Positional Accuracy
- 2.4.1. Horizontal Positional Accuracy
  - 2.4.1.1. Horizontal Positional Accuracy Report
  - 2.4.1.2. Quantitative Horizontal Positional Accuracy Assessment
    - 2.4.1.2.1. Horizontal Positional Accuracy Value
    - 2.4.1.2.2. Horizontal Positional Accuracy Explanation
- 2.4.2. Vertical Positional Accuracy
  - 2.4.2.1. Vertical Positional Accuracy Report
  - 2.4.2.2. Quantitative Vertical Positional Accuracy Assessment
    - 2.4.2.2.1. Vertical Positional Accuracy Value
    - 2.4.2.2.2. Vertical Positional Accuracy Explanation

2.5. Lineage

2.5.1. Source Information
- 2.5.1.1. Source Citation
- 8. Citation Information
- 2.5.1.2. Source Scale Denominator
- 2.5.1.3. Type of Source Media
- 2.5.1.4. Source Time Period of Content
  - 9. Time Period Information
  - 2.5.1.4.1. Source Currentness Information
- 2.5.1.5. Source Citation Abbreviation
- 2.5.1.6. Source Contribution

2.5.2. Process Step
- 2.5.2.1. Process Description
- 2.5.2.2. Source Used Citation Abbreviation
- 2.5.2.3. Process Date
- 2.5.2.4. Process Time
- 2.5.2.5. Source Produced Citation Abbreviation
- 2.5.2.6. Process Contact
- 10. Contact Information

2.6. Cloud Cover

3. Spatial Data Organization Information

3.1. Indirect Spatial Reference

3.2. Direct Spatial Reference Method

3.3. Point and Vector Object Information
- 3.3.1. SDTS Terms Description
  - 3.3.1.1. SDTS Point and Vector Object Type
  - 3.3.1.2. Point and Vector Object Count
- 3.3.2. VPF Terms Description
  - 3.3.2.1. VPF Topology Level
  - 3.3.2.2. VPF Point and Vector Object Type
  - 3.3.1.2. Point and Vector Object Count

3.4. Raster Object Information
- 3.4.1. Raster Object Type

continue on next figure

- n. Constant Element
- n. Constant Element by Data Content
- n. Constant Element by Convention
- n. Variable Element

Figure 6.27 Classification of constants and variable element of the American standard for metadata. According to Hedorfer and Bianchin (1998).

3.4.2. Row Count
3.4.3. Column Count
3.4.4. Vertical Count

4. Spatial Reference Information
4.1. Horizontal Coordinate System Definition
4.1.1. Geographic
4.1.1.1. Latitude Resolution
4.1.1.2. Longitude Resolution
4.1.1.3. Geographic Coordinate Units

4.1.2. Planar
4.1.2.1. Map Projection
4.1.2.1.1. Map Projection Name
4.1.2.1.2. Map Projection Parameters (see CSDGMD)
4.1.2.2. Grid Coordinate System
4.1.2.2.1. Grid Coordinate System Name
4.1.2.2.2. Grid Coordinate System Param. (see CSDGMD)

4.1.2.3. Local Planar
4.1.2.3.1. Local Planar Description
4.1.2.3.2. Local Planar Georeference Information

4.1.2.4. Planar Coordinate Information
4.1.2.4.1. Planar Coordinate Encoding Method
4.1.2.4.2. Coordinate Representation
4.1.2.4.2.1. Abscissa Resolution
4.1.2.4.2.2. Ordinate Resolution
4.1.2.4.3. Distance and Bearing Representation
4.1.2.4.3.1. Distance Resolution
4.1.2.4.3.2. Bearing Resolution
4.1.2.4.3.3. Bearing Units
4.1.2.4.3.4. Bearing Reference Direction
4.1.2.4.3.5. Bearing Reference Meridian
4.1.2.4.4. Planar Distance Units

4.1.3. Local
4.1.3.1. Local Description
4.1.3.2. Local Georeference Information

4.1.4. Geodetic Model
4.1.4.1. Horizontal Datum Name
4.1.4.2. Ellipsoid Name
4.1.4.3. Semi-major Axis

4.1.4.4. Denominator of Flattening Ratio

4.2. Vertical Coordinate System
4.2.1. Altitude System Definition
4.2.1.1. Altitude Datum Name
4.2.1.2. Altitude Resolution
4.2.1.3. Altitude Distance Units
4.2.1.4. Altitude Encoding Method
4.2.2. Depth System Definition
4.2.2.1. Depth Datum Name
4.2.2.2. Depth Resolution
4.2.2.3. Depth Distance Units
4.2.2.4. Depth Encoding Method

5. Entity and Attribute Information
5.1. Detailed Description
5.1.1. Entity Type
5.1.1.1. Entity Type Label
5.1.1.2. Entity Type Definition
5.1.1.3. Entity Type Definition Source

5.1.2. Attribute
5.1.2.1. Attribute Label
5.1.2.2. Attribute Definition
5.1.2.3. Attribute Definition Source
5.1.2.4. Attribute Domain Values
5.1.2.4.1. Enumerated Domain
5.1.2.4.1.1. Enumerated Domain Value
5.1.2.4.1.2. Enumerated Domain Value Definition
5.1.2.4.1.3. Enumerated Domain Value Def. Source
5.1.2. Attribute
5.1.2.4.2. Range Domain
5.1.2.4.2.1. Range Domain Minimum
5.1.2.4.2.2. Range Domain Maximum
5.1.2. Attribute
5.1.2.4.3. Codeset Domain
5.1.2.4.3.1. Codeset Name
5.1.2.4.3.2. Codeset Source
5.1.2.4.4. Unrepresentable Domain
5.1.2.5. Attribute Units of Measure
5.1.2.6. Attribute Measurement Resolution

5.1.2.7. Beginning Date of Attribute Values
5.1.2.8. Ending Date of Attribute Values
5.1.2.9.1. Attribute Value Accuracy
5.1.2.9.2. Attribute Value Accuracy Explanation
5.1.2.10. Attribute Measurement Frequency

5.2. Overview Description
5.2.1. Entity and Attribute Overview
5.2.2. Entity and Attribute Detail Citation

6. Distribution Information
6.1. Distributor
10. Contact Information
6.2. Resource Description
6.3. Distribution Liability
6.4. Standard Order Process
6.4.1. Non-digital Form
6.4.2. Digital Form
6.4.2.1. Digital Transfer Information
6.4.2.1.1. Format Name
6.4.2.1.2. Format Version Number
6.4.2.1.3. Format Version Date
6.4.2.1.4. Format Specification
6.4.2.1.5. Format Information Content
6.4.2.1.6. File Decompression Technique
6.4.2.1.7. Transfer Size
6.4.2.2. Digital Transfer Options
6.4.2.2.1. Online Option
6.4.2.2.1.1. Computer Contact Information
6.4.2.2.1.1.1. Network Resource Name
6.4.2.2.1.1.2. Dialup Instructions
6.4.2.2.1.2.1. Lowest BPS
6.4.2.2.1.2.2. Highest BPS
6.4.2.2.1.2.3. Number DataBits

continue on next figure

n. Constant Element
n. Constant Element by Data Content
n. Constant Element by Convertion
n. Variable Element

Figure 6.27 cont'd.

6.4.2.2.1.1.2.4. Number Stoptbis
6.4.2.2.1.1.2.5. Parity
6.4.2.2.1.1.2.6. Compression Support
6.4.2.2.1.1.2.7. Dialup Telephone
6.4.2.2.1.1.2.8. Dialup File Name
6.4.2.2.1.2. Access Instructions
6.4.2.2.1.3. Online Computer and Operating System
6.4.2.2.2. Offline Option
6.4.2.2.2.1. Offline Media
6.4.2.2.2.2. Recording Capacity
6.4.2.2.2.2.1. Recording Density
6.4.2.2.2.2.2. Recording Density Units
6.4.2.2.2.3. Recording Format
6.4.2.2.2.4. Compatibility Information

6.4.3. Fees
6.4.4. Ordering Instructions
6.4.5. Turnaround
6.5. Custom Order Process
6.6. Technical Prerequisites
6.7. Available Time-Period
9. Time Period Information

7. Metadata Reference Information
7.1. Metadata Date
7.2. Metadata Review Date
7.3. Metadata Future Review Date
7.4. Metadata Contact
10. Contact Information
7.5. Metadata Standard Name
7.6. Metadata Standard Version
7.7. Metadata Time Convention
7.8. Metadata Access Constraints
7.9. Metadata Use Constraints
7.10. Metadata Security Information
7.10.1. Metadata Security Classification System
7.10.2. Metadata Security Classification
7.10.3. Metadata Security Handling Description

8. Citation Information
8.1. Originator
8.2. Publication Date
8.3. Publication Time
8.4. Title
8.5. Edition
8.6. Geospatial Data Presentation Form
8.9. Series Information
8.9.1. Series Name
8.9.2. Issue Identification
8.10. Publication Information
8.10.1. Publication Place
8.10.2. Publisher
8.11. Other Citation Details
8.12 Online Linkage
8.13. Larger Work Citation
8. Citation Information

9. Time Period Information
9.1. Single Date/Time
9.1.1. Calendar Date
9.1.1. Time of Day
9.2. Multiple Dates/Times
9.1.1. Calendar Date
9.1.1. Time of Day
9.3. Range of Dates/Times
9.3.1. Beginning Date
9.3.2. Beginning Time
9.3.3. Ending Date
9.3.4. Ending Time

10. Contact Information
10.1. Contact Person Primary
10.1.1. Contact Person
10.1.1. Contact Organization
10.2. Contact Organization Primary
10.1.1. Contact Person
10.1.1. Contact Organization
10.3. Contact Position

10.4. Contact Address
10.4.1 Address Type
10.4.2 Address
10.4.3. City
10.4.4. State or Provence
10.4.5. Postal Code
10.4.6. Country
10.5. Contact Voice Telephone
10.6. Contact TDD/TTY Telephone
10.7. Contact Facsimile Telephone
10.8. Contact Electronic Mail Address
10.9. Hours of Service
10.10. Contact Instructions

n. Constant Element
n. Constant Element by Data Content
n. Constant Element by Convention
n. Variable Element

Figure 6.27 cont'd.

6.8.4 *Implementing metainformation with relational databases*

Recently, Qiu and Hunter (1999) and Gan and Shi (1999) have proposed new methods for storing metainformation based on a relational model. Starting from those works, we can propose the following:

```
THEME ( Theme-ID, Themename, Description)
QUA-THEME (Theme-ID, lineage, position, attribute,
logical-consistency, completeness)
LAYER ( Layer-ID, Theme-ID)
QUA-LAYER (Layer-ID, lineage, position, attribute,
logical-consistency, completeness)
FEATURE-CLASS ( Fclass-ID, Fclass-ID, Feature-code,
Fclassname, Layer-ID)
QUA-CLASS (Test-ID, Fclass, position-method,
position-check-date, position-check-result,
completeness-method, completeness-check-date,
position-check-result, logical-consistency-method,
logical-consistency-check-date, logical-
consistency-check-result)
```

6.8.5 *List of spatial metainformation systems*

Some spatial metainformation systems can be found at the following addresses:

- European Spatial Infastructure (ESMI): http://www.megrin.org/gddd/ esmi.htm
- European Wide Service Exchange (EWSE): http://ewse.ceo.org
- Geographical Data Description Directory (GDDD): http://www. megrin.org/gddd
- Multilingual Catalogue of Data Sources (CDS): htpp://www. mu.niedersachsen.de/cds
- etc.

6.9 Conclusions

As a conclusion to this chapter let me say that if now we know how to model data, we still have problem for modelling and manipulating knowledge. Several tracks have been explored in the past decades, but after billions of studies, most of them were discovered as wrong tracks. But the track of metainformation seems to be very promising, even if it is a little boring to maintain metadata daily; one interesting idea is to nominate a metadata manager along with the data custodian and the GIS administrator. In very small organisations, the same person can perform these three tasks, but in huge organisations, several persons must be in charge of these tasks.

In our domain, I think that we still have difficulties modelling urban, and especially location-based knowledge. Even though several years ago, I co-authored (Laurini and Milleret-Raffort 1989) a book on knowledge engineering, I still believe that the solution of an efficient way of modelling spatial knowledge has not been approached.

Concerning spatial metadata, the concept of a clearinghouse, that is to say regrouping and qualifying all spatial information by means of metadata is very interesting work. For a nice example in Illinois, USA, please visit their site: http://www.isgs.uiuc.edu/nsdihome/ISGSindex.html; and for roads, http://www.inhs.uiuc.edu/nsdihome/outmeta/roads.html.

Finally let me present this interesting drawing (Figure 6.28) in which a gradation is explained from data to wisdom through information and knowledge.

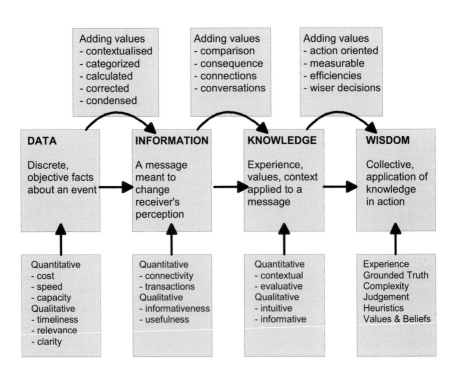

Figure 6.28 The knowledge progression, from data to wisdom. From Sena and Shani (1999).

7

VISUALISATION FOR DISPLAYING AND ACCESSING URBAN INFORMATION

Visualisation is much more than cartography. Traditional urban planners are using cartography quite a lot, passing gradually from paper cartography to computer cartography. But now computers can go beyond conventional cartography, especially by using the capabilities of animation, augmented reality, real time and so on. It is very possible to affirm that a new kind of cartography is emerging (Müller and Laurini 1997), the main characteristics of which are as follows:

- exclusive use of computers, no more paper-based cartography
- screens, and especially high-resolution screens in use
- **dynamic aspects,** or in other words animated cartography, especially in real time
- **active aspects,** that is to say the connections with hypermaps or hyperlinks.

Another family of characteristics comes from the finality of maps. For centuries maps were considered as a medium both for storing and representing geographic and spatial information. The storing aspect is now dedicated to GIS databases, the representation for conventional mapping even if it was based on computers. But computer screens also allow maps to be used as entry into any information systems, often known as visual interfaces. In the previous chapter on hypermaps, we saw maps as entry windows to documents. So, the role of maps, and more generally of visualisation techniques as inputs to computer systems is increasingly important. In this chapter, we will also deal with these aspects and it will be organised accordingly.

As previouly said, visualisation is not only cartographic output: it can also be used as input to databases and also for navigating through the databases. Indeed, several novel systems for access to and navigating in the databases have been proposed, and some of them appear very interesting for urban information systems. In this chapter, first will be presented the main characteristics for passing from conventional to computer-based cartography. Then new tools for visualisation as output will be examined, and finally some

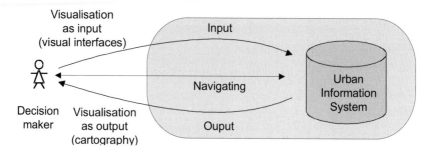

Figure 7.1 Visualisation is not only cartography; visualisation can also help the decision-maker as visual interfaces to access data, to navigate in the database, to display the results and to run applications.

elements will be given for understanding visual interfaces for accessing databases, and especially urban databases.

7.1   Generalities[1]

For years, visualisation in urban planning was essentially performed for the delivering of maps to the decision maker, mainly in the form of paper-sheets. But as shown in Figure 7.1, visualisation can help the decision maker not only as outputs of some computer applications, but also as input visual interfaces in order to access the database and run applications.

Before developing input and output visualisations, let us present very rapidly the novelties regarding them.

7.1.1   Cartographic visualisation

These two terms seem tautological since cartographic activities always lead to a graphic representation, hence a visual product. In fact the expression 'cartographic visualisation', which has already been used by many others (Taylor 1994), suggests the association between a traditional discipline and a new investigation tool. This discipline relies on the systematic exploitation of new display technologies, such as computer graphic images, numerical image processing, holography, multidimensional animations, visual simulations and virtual reality. Theoretical thinking and scientific analysis are then enhanced by the possibility of 'seeing' and studying phenomena which were previously hidden or not clearly visible to the human eye. In fact visualisation does not only mean

---

1   The first part of this chapter (Section 7.1) is derived from: Müller and Laurini, *La cartographie de l'an 2000*, Revue Internationale de Géomatique, Volume 7, 1, 1997 pp. 87–106.

improved communication with the help of our visual organs, it also defines a new way of 'dialoguing' with an image, including the possibility to explore, manipulate, discover and communicate spatial and multidimensional relations through ephemeral and dynamic representations; Ephemeral, to the extent that an interactive dialogue between operator and system affords on-line manipulation and processing of an image (changes in colour, perspective and symbolism, browsing through various representations and layers in the realm of investigation, etc.) in order to identify pertinent facts such as trends and anomalies. 'Dynamic', to the extent that animation, as we know it from daily exposure to television, affords the discovery and analysis of temporal processes (growth, modification of evolution, temporal thresholds etc.). Visualisation also expresses a paradigm shift in the function of maps, since it emphasises the notion of a database view of maps (Peterson 1995), when maps are used to query against a database. Hence 'visualisation' gives a new dimension to cartography, since the capability to explore and manipulate multidimensional scientific data interactively (MacEachren 1994) is added to the traditional role of simple, one-way, map-reader communication.

Bertin (1967, 1983), in his graphic semiology, proposed several variables to represent graphically the information (visual variables):

- differences in size, that is to say how large a point, a line or a zone is
- differences in lightness or grey value
- differences in grain, that is to say the texture
- differences in colour hue
- differences in orientation
- and differences in shape.

Now, with computers, we can add other possibilities:

- animation governed either in real time or differed time; time can be accelerated or decelerated; if needed, one can replay an animated map
- simulated flights through the landscape or the cityscape
- links to other multimedia information, for instance to a written document, photos, some auditory information, a movie, etc.; sometimes they are called hyperlinks.

Back to variables, in addition to visual variables, some authors (Kraak and MacEachren 1994 and Kraak and Ormeling 1996) have proposed dynamic variables such as:

- display time
- duration
- frequency
- order
- rate of change
- synchronisation.

Let us examine some aspects.

7.1.2   Real time visual animation

New applications such as the control of pollution or the prevention and the management of hazards require the use of animated maps. This need is obvious in monitoring systems as well as in teams for crisis management where rapid decisions must be taken, and in which space and location play an important role (see Chapter 11). These maps are characterised by the following elements:

- information is captured by sensors and sent according to various telecommunication techniques to a central computer which is in charge of interpretation and visualisation;
- the speed of data acquisition varies according to the applications, spanning from one second or perhaps less, to several days;
- from each information sensor, procedures are triggered and visualisation is performed with a more or less sophisticated scheme;
- visualisation is thus animated in real time;
- if necessary, one can 'replay' the past in order to capture more easily the evolution;
- in some cases, it can be of interest to accelerate or decelerate the time when carrying out for instance what-if simulations.

Among the pieces of information received (sometimes several thousand chunks per second), some are more important than others; a sophisticated system must be constructed to allow the automatic selection of important variables and select a graphic semiology which gives the possibility for an operator to understand the situation in a condensed way (e.g. providing a synoptic view). Such systems already exist in the form of dedicated expert systems, especially for the command and control of nuclear centres, without cartographic features. One may consider, in the near future, the support of huge-format screens, dynamic cartographic views combining various elements from graphic semiology, whenever a spatial dimension is involved. We think in particular of applications such as traffic control on highways, monitoring of flooding and management of corrective actions in crisis cases, etc. (Figure 7.2).

7.1.3 Multimedia variables

As previously said, visual variables, such as defined by Bertin (1967), constitute a good framework from which rules can be built for the appropriate use of symbols in cartography. There arises the question of whether we could also formalise the rules which would be applicable when using dynamic and auditory variables. It is advantageous to draw a parallel between animation and sounds since they both have a temporal dimension. Although sounds and

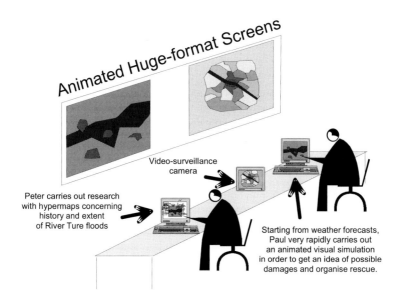

Animated Huge-format Screens

Video-surveillance camera

Peter carries out research with hypermaps concerning history and extent of River Ture floods

Starting from weather forecasts, Paul very rapidly carries out an animated visual simulation in order to get an idea of possible damages and organise rescue.

Figure 7.2 Examples of control centre working with a real-time visualisation system including animated visual simulation.

auditory information may be a useful complement to the presentation of static documents (i.e. spoken comments of a map or a statistical table), its application is most particularly helpful when doing animation (both messages run parallel). **A semiology of auditory variables** was suggested by Krygier (1994). They include the location, strength, pitch, register, timbre, duration, sequential order, and attack. The utilisation of these variables to 'represent' nominal, ordinal, and ratio data is still a matter of speculation, although Krygier (1994) suggests some application criteria. (See also Gomes *et al.* 1999).

Sounds may be considered in two ways: as spatial information or as metainformation. For example, in the mapping of urban noise, sound may be considered a spatial variable (whose representation could be performed through the sound itself; in such a case urban noise is both reality and representation, a situation which would be difficult to emulate with temperature or precipitation!). We can imagine a system which enables the retrieval urban noise in every place at every time, providing the data are available (clicking with the mouse or tracing with the mouse over the map space would trigger a noise discretely or continuously). Another interesting problem is the representation of urban noise with conventional cartographic methods, since it is quantitative as well as qualitative, and is continuous over space and time. Isoline maps (iso-sound lines) can be used (Figure 7.3), but they provide only partial solutions (only parts of the sound aspects are depicted). Hence the necessity of combining graphics with other media in order to

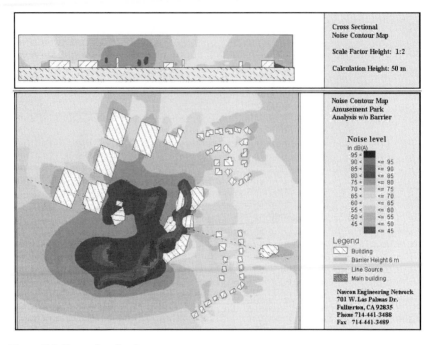

Figure 7.3 Example of noise contour map.
Source: Navcon Engineering Network http://www.navcon.com/splana7.htm #Cross Sectional Map Software: SoundPLAN Version 5.0.

communicate the diversity of sounds, and their psychological and physiological effects (pleasant versus unpleasant, tiresome or troublesome, etc.).

Another display of the location where the measures were made is based on balloons (Figure 7.4a). Finally also auditory tape recording (typically from one to several minutes) can be made and located (Figure 7.4b). Also smiling faces can be used to represent a global evaluation (Figure 7.4c). In the latter case, by clicking on the signal, the computer can emit the sound and the user can hear it.

7.1.4 Animated visual simulation

Animation is naturally well suited for the representation of dynamic processes. Take the case of flooding. We are presently able to carry the numerical simulation of water flow. One could present the results in the form of long page listings, but a visual simulation is obviously more pleasant and much more convincing.

Generalities

Several types of visual simulations may be realised including:

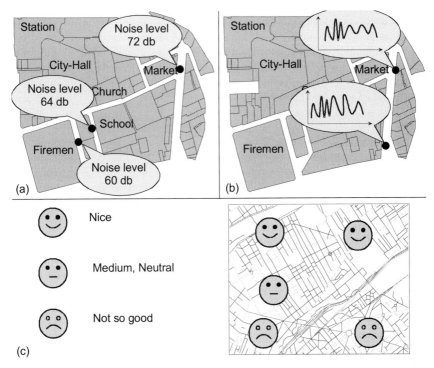

Figure 7.4 Example of visualisation of auditory information. (a) Using balloons to locate where the measures were made. (b) Location of the tape recorded auditory signal. (c) Using smiling faces.
From Laurini and Servigne (1999).

- A **realistic visual simulation** is computationally very time-consuming, not only for the numerical simulation of phenomena, but also for the rendering of light effects; here the symbolism is reduced, and one tends to use more the possibilities offered by artificial movies; a possibility is to combine animation with augmented reality.
- A **symbolic visual simulation** does not imply a realistic rendering, it relies on the symbolic, i.e. a semiology which puts forward some striking points, some essential variables, etc. Here lies the foundation for a new animated cartography.

Some applications require slowing down or speeding up time; this change in speed leads to the notion of temporal scale. For example, in the visual simulation of forest growth, one would like to speed up the time, whereas in nuclear physics, one needs a slower time sequence. In the case of geoprocessing there seems to be no application requiring slower motions; rather, one needs usually to accelerate time. This acceleration may occur at totally different rates. Think for example of the visual simulation of demographic evolution or the monitoring of road traffic. Furthermore, we must make a distinction between real time and differed time. With **real-time visual simulation**, we mean a simulation for which the computational results are directly displayed (output) according to the rhythm of computations or according to the capture of remote data (sensors). **Differed time visual simulation** is used when lengthy processing is carried out, for example during the night. Afterwards, when the computations are finished, the results are displayed with the appropriate speed. Due to the improved speed of the computers and the possible spread of parallel machines for the processing of complex phenomena, differing visualisation of simulation results will probably tend to disappear and be replaced by real-time simulations.

After having made the above distinction, we still face the problem of symbolisation in animated cartography. For sounds, one must establish a typology of dynamic variables (duration, mutation, sequencing, synchronisation, speed, acceleration, etc.) which could be used to visualise data measured according to a nominal, ordinal, or interval scale. Much research is needed about the way dynamic variables may be used for visualisation purposes. How do you create a legend including dynamic variables or auditory variables which could effectively define the nature or the magnitude of the objects to be represented? On the other hand, the sequential nature of the sound or animation presents a problem in terms of knowledge acquisition or memorisation (contrary to graphics, acoustic signs or animated signs are ephemeral). These difficulties, added to those of traditional communication, could disorientate the potential user. From a more optimistic point of view, one could think that multimedia communication could help traditional graphic communication due to possible redundancies with converging effects (a sound may help to reinforce the identity of a graphic symbol, as for example the chime of a bell when clicking on a symbol suggesting a 'church' on the map; other less pictorial, more abstract symbols could also be identified by sound).

Applications in urban planning

For several urban applications, animation is a very interesting way to visualise information. Among examples let us speak about traffic visualisation. Another possibility is for noise visual representation (Figure 7.5) from Kang and Servigne (1999). See also http://yerkes.mit.edu/shiffer/MMGIS/noisenic.html for another example of noise animation made by Shiffer (1999).

To conclude this short introduction to sound and animated visual simulation, let us mention that we need to take a number of artefacts into account. The best known is that of flowing water which 'climbs back'. Indeed, if we combine the periodicity of the computations with the periodicity of the screen sweeping, stroboscopic effects may arise analogue to those of the wheels of a coach which roll backward (as in old western movies). Thus, in the setup of animated visual simulation systems, we must be careful to avoid this

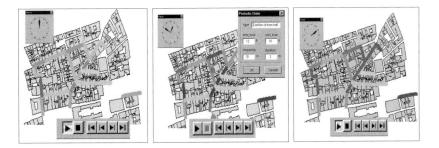

Figure 7.5 Noise animation.
From Kang and Servigne (1999).

kind of problem which may lead to totally wrong interpretations, although the programs were originally correct in their conception.

7.1.5 Metaphors

According to the Merriam-Webster dictionary, a metaphor is 'a figure of speech in which a word or phrase literally denoting one kind of object or idea is used in place of another to suggest a likeness or analogy between them (as in drowning in money)'. In computing, it means using an analogy for designing a system, or especially its human interface. For instance, the metaphor of an office first proposed by Apple, is now commonly used in interface design.

Recently, considering the evolution of technology (especially Internet technology) and its incidence on cartography Cartwright (1999) has proposed extending the map metaphor using delivered multimedia. So, in addition to maps, he gave nine metaphors for displaying geo information. These metaphors are seen to offer access genres that are complementary to the use of maps, and are, thus, intended to be used in conjunction with maps:

- **storyteller**, by which a user is being told about the geography of a designated area, for instance its evolution;
- **navigator**, by offering a tool to assist users in finding where information is located; more particularly this metaphor will give support for users who were not good at interpreting abstract models, including maps;
- **guide**, when assuming that the user has, neither prior knowledge of the area being portrayed, nor the ability to effectively navigate through individual scenes;
- **sage**, as a metaphor suggests access to expert advice or information that provides support for decision-making or information appreciation; by using hot links, the user can get immediate answers to his interrogations; this metaphor is similar to clickable hypermaps;

- **data store**, by linking to other information about the area under current investigation, without needing to display all of this information;
- **fact book**, by enabling access to a plethora of facts, both about an area under study and other areas about which comparison might have to be made;
- **gameplayer**, by offering things such as a map-building game that allows users to learn the grammar of mapping; the goal is not to play the game, but to use gaming skills to explore geographic information;
- **theatre**, which is based on a stage, players and a script; the stage is the three-dimensional space (plus time), the players are the things that occur on the stage, or the elements of the landscape, and the script can be written either by the product author, or interactively by the user;
- **toolbox**, by offering a set of tools to users, perhaps to make decisions or to explore information.

These metaphors, and maybe others will be used throughout this chapter in order to organise information and applications by potential users. A key-element often required is to use only a unique consistent metaphor during an application. When it is necessary to change metaphors, the user must be informed about this new way of structuring.

7.1.6   Present technical barriers to overcome

According to Gahegan (1999), the science of visualisation contains elements from many disciplines, including computer graphics, cartography, human perception, and knowledge engineering. In order to reach successful application, it must overcome four barriers that are technological, perceptual, computational and geographical, namely:

- rendering speed: technological barriers affecting the interactive display and manipulation of large datasets
- visual combination effects (seeing the wood for the trees): problems to be overcome when displaying many layers, themes or variables simultaneously
- complexity of the mapping between geographical datasets and the visual domain (finding a needle in a haystack)
- and the vast range of possibilities that a visualisation environment provides; the orientation of the user into a visualised scene or virtual world and measures of the effectiveness of the scene or world as a problem-solving tool (where am I in the virtual world?).

7.2   Visualisation as output

Visualising urban planning information is much more than cartography, especially with the possibilities of virtual reality and augmented reality. Among possibilities, let us explore 3D rendering of cities and photo-based

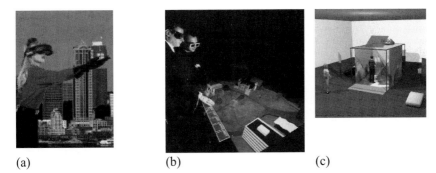

(a)   (b)   (c)

Figure 7.6 Different systems for virtual reality. (a) Example of an head-mounted display (HMD) model CyberEye™ the General Reality Company. Source: http://www.ireality.com/. General Reality Company is a subsidiary of iReality (b) virtual workbench. Source: http://viswiz.gmd.de/IMF/rw.html. Published with permission. (c) surround projection in a cave.
Source: http://www.cica.indiana.edu/ (Verbree *et al.* 1999) Eric Wernet, © 1999, Indiana University. See also Figure 9.4.

navigation throughout a city. In urban planning, 3D rendering means four different things:

1   Realistic rendering, that is to say giving a 3D view of the actual city; it corresponds to a visualisation, often simplified or formalised of what is now in the city (an illustration is given in Figure 9.6)
2   Urban project rendering, or more exactly visually representing new development as exemplified in Figure 9.7; this kind of rendering can be used during the design of a new urban project, perhaps to visualise different planning alternatives
3   Prescriptive rendering or the visualisation of effects of certain planning rules which will be detailed later in this section
4   Symbolic rendering in which the scope is not to represent existing or planned 3D objects but after having selected some graphic semiology to represent some attributes for 3D objects as exemplified in Figure 7.6.

7.2.1   *Virtual reality and urban planning*

Virtual reality (VR) may best be defined as the wide-field presentation of computer-generated, multi-sensory information which tracks a user in real time. According to Kruijff (1998), different types of immersions can be used.

Full immersion VR systems deliver the highest sense of presence – almost every functional aspect in Figure 7.6 can theoretically be included, thereby supplying users with auditory, visual and force feedback sensations. To reach full immersion, this kind of VR system always makes use of a head-coupled viewing device like a Head Mounted Display (HMD) (Figure 7.6a),[2] creating a

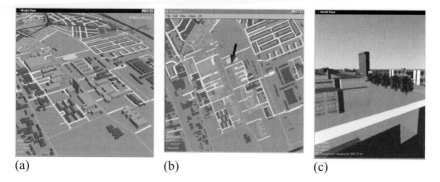

(a)   (b)   (c)

Figure 7.7 Different view nodes visualising a part of the centre of the city of Utrecht in the Netherlands.
Source: http://karma.geo.tudelft.nl (Verbree *et al*. 1999), published with permission.

single user experience. The HMD supports the user with images via a helmet with two displays connected to an optical system. The displays provide the user with only a limited field of view using relatively low resolutions, but have a 360° field of view. Actual HMDs are not very ergonomical because of their weight and the amount of cables which need to be connected to the helmet.

Semi-immersion VR systems use screens to display the images. Systems vary from single-screen installations to room-like installations like the cave[3] (Figure 7.6c) in which images can be displayed on multiple walls, enabling an extreme wide field of view. In addition, the word cave makes a reference to 'The Simile of the Cave' found in Plato's *Republic*, in which the philosopher explores the ideas of perception, reality, and illusion. Plato used the analogy of a person facing the back of a cave alive with shadows that are his/her only basis for ideas of what real objects are. In VR caves, users are wearing some special glasses. The quality of the displayed images is very high – the field of view is wide and the high resolution delivers realistic images. Theoretically, semi-immersion systems can use the same control devices, including haptic, force and auditory feedback, as full immersion systems.

2   Originally by General Reality Company purchased by iReality Company.
3   US Argonne National Laboratory's *Cave Automatic Virtual Environment* (CAVE), http://www._fp.mcs.anl.gov/fl/research/facilities/index.htm one of four such sophisticated virtual reality facilities in the world, lets scientists see, touch, hear and manipulate data. Designers can use the facility for activities such as creating and testing models of commercial boilers and visualising a positron beam as it travels through and is influenced by high-powered magnets. The CAVE is a 10-foot-square room composed of projection screens. In a darkened lab, projectors and mirrors overlap two images on each screen. Inside the CAVE, researchers wear stereoscopic glasses to turn the projections into hologram-like images. Six to 10 people can stand in the CAVE and view the projection, while one person with a headset and joystick-like computer wand controls the simulation's perspective. The CAVE was developed by the Electronic Visualization Lab at the University of Illinois at Chicago.

In practice, both full and semi-immersion systems seldom make use of all the available technology. A main advantage of semi-immersion systems is that co-operation is fairly easy to achieve – multiple participants can immerse in the virtual environment and at the same time see each other through the glasses. A special kind of semi-immersion system is the Responsive Workbench, developed by GMD (see Figure 7.6b). Users of the Responsive Workbench view stereoscopic images which are projected onto a tabletop via a projector-and-mirrors system. The used tabletop metaphor is valuable for urban planning applications.

In addition, Ernst Kruijff (1998) has split up architectural virtual reality applications in several very different areas:

- **VR Modelling.** Tools which are specifically meant for assembling three-dimensional scenes, especially by using VRML (Virtual Reality Markup Language). Because it is relatively hard to model very detailed scenes with these tools, they are only applicable in the beginning of the architectural design process.
- **VR Walkthroughs.** Walkthroughs are the most frequently used VR applications in the architectural design process up to now. Two kinds of walkthroughs can be defined. The first kind of walkthrough is the more traditional one in which one can view a building by walking through it via a predefined path or by using an input device to explore the building personally. The second kind is what I would call the interactive walkthrough, in which interaction with the digital model of the building is possible. Interactive walkthroughs can be a valuable tool for evaluation, because one is also able to change certain parts of the building in (almost) real time.
- **Virtual Archaeology.** VR is an excellent tool to bring buildings of the past to life again, enabling the sharing of discoveries and insights between researchers. Famous examples are the Virtual Pompeii by SIMLAB and the reconstructions of the Basilica of St. Peter's and the City of Giotto (Bertol 1997).
- **Urban Planning.** VR applications for urban planning closely resemble VR walkthroughs, in a way that they are also used for the exploration of physical space. In the case of urban planning, the focus is on place and context of the buildings in an urban environment, not at the buildings themselves. See Figure 7.7 for examples in the Dutch city of Utrecht, kindly provided by Edward Verbree (Verbree and Verzijl 1998, Verbree *et al.* 1998, 1999).

These elements will be used especially in the Public Participation Chapter (Section 9.3).

In augmented reality, an attempt is made to add some other information to the reality, giving a sort of symbolic rendering. See for instance Figure 7.8 in which buildings are not represented realistically, but according to their use.

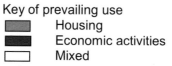

Key of prevailing use
- Housing
- Economic activities
- Mixed

Figure 7.8 Symbolic rendering: Example of 3D visualisation together with activities. After Vico and Rossi-Doria (1997).

7.2.2 Prescriptive rendering

In all countries, planning laws allow landowners to build a house or to extend it, perhaps with additional levels. For instance, in France, a very important planning parameter is floorspace ratio. But, with the same value of this parameter, land built forms will be very different. An example illustrates this case in Figure 7.9 with a single rule 'Prescriptive maximum floorspace ratio = 50%'. As can be seen, this single rule can generate different kinds of cityscape. Figure 7.10a gives some of the various possibilities for the same plot. And the successive Figures 7.10b, c and d, depict some possible cityscapes generated by this rule (constructibility). More generally, for a single plot, different kinds of rules can apply, not only those regarding the floorspace ratio; indeed other parameters can influence, such as the minimum distance to street, to the neighbours, and so on.

A more complex example is given in Figure 7.10 (from Vico and Rossi-Doria, 1997, Laurini and Vico, 1999) in which one can see the possible effects of some rules on a streetscape in an Italian city.

7.2.3 CityView

A system named CityView was developed by the French company SNV (Société de Numérisation de Ville). The goal is to store the façades of all buildings in order to display each side of a street. For details see http://www.snv.fr. An example is given in Figure 7.11. Due to the 'wheel' button located bottom-left, it is possible to go forward or backward in the city, that is to say to see it as a driver driving his car in Paris. It is possible to enter the system, either by a street address, or by giving the name of a building (for instance, the Eiffel Tower).

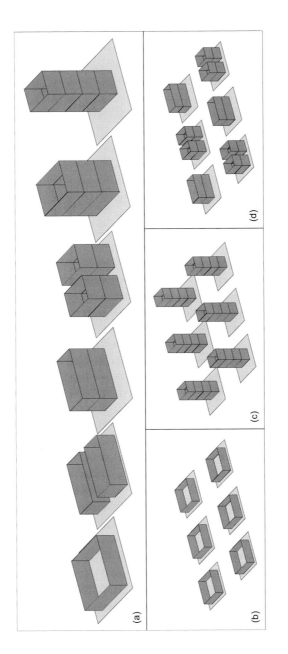

Figure 7.9 Several possibilities of land use built forms derived from the rule 'maximum prescriptive floorspace ratio = 50 %'. (a) Some of the possibilities. (b) Example with one-level building per parcel. (c) A tower per parcel. (d) Mixed area.

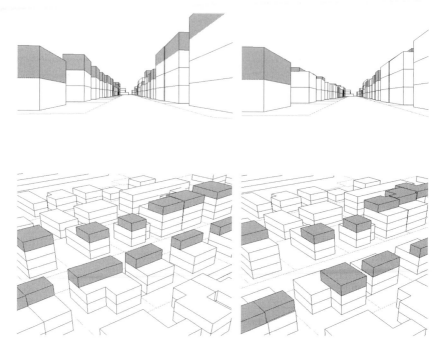

Figure 7.10 Example of land built forms from different viewpoints and different planning rules. Upper row, pedestrian view; lower row, helicopter view. Left column, max hypothesis: buildings extension lining street line; right column, likely hypothesis: make uniform the building height.
According to Vico and Rossi-Doria (1997), Laurini and Vico (1999).

7.3 Visual interfaces for accessing, and navigating in, urban data

As previously said, visualisation can also be used for accessing data and navigating through them. In this section, several examples will be examined. First, some global visualisation systems will be presented, then some for navigating such as the information space and the information city. Finally an example in flood history and prevention will be detailed.

7.3.1 Global visualisation systems

Ben Shneidermann (1992, 1993) and Card *et al.* (1999) from the University of Maryland, College Park and his team created various methods for the global visualisation of database contents. In contrast with the usual case of databases available from SQL where the user does not even know the content of the base he is querying, global visualisation systems are based upon various global sequential facilities such as **overview**, **zoom and filter**, and **details-on-demand**. Let us examine the key steps of this procedure:

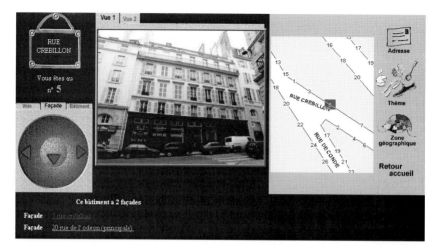

Figure 7.11 Example of an interface of a system for storing, retrieving and displaying building façades.
Source: http://www.snv.fr. Used with kind permission of Société Numérisation de Ville.

- Once a database is opened, a global visualisation of the content is made available, that is, all objects of the database are symbolised on the screen; two techniques are proposed: (a) the **starfield** which consists of representing each object by a point in two-dimensional space (those two dimensions are carefully selected), and (b) the **space-filling treemap**, (Johnson and Shneiderman, 1991) (Venn diagrams arranged in branching trees) which consists of regrouping the objects in hierarchical categories;
- Furthermore, objects of interest are iteratively selected through zooming and filtering until the desired objects are retrieved;
- Finally, once the searched objects have been selected, their respective attributes are presented to the user.

Back to SQL, we can see SQL as a sequence (SELECT.. FROM.. WHERE..) in which, are first mentioned the details, then the list of tables and finally some criteria. In other words, the SQL sequence can be seen as *Details, Overview and Filtering*, the order of which is not very obvious.

Starfield

When dealing with a single class of objects, the principal idea of the starfield technique is to display all instances of this class in order to access the database. So, the starfield will be displayed taking the characteristics of screens (2D and colours) into account. Namely three attributes can be selected to represent first the 2D space, one for the x-axis, one for the y-axis, and the third is represented through colours as exemplified in Figure 7.12.

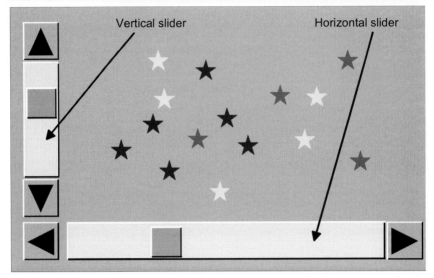

Figure 7.12 Principles of the starfield approach, one attribute for the *x*-axis, one for the *y*-axis, and a third one as colour for star points.

By using both horizontal and vertical sliders, all the data space can be scanned. Some widgets can be added to assign colours to categories, to modifiy the types of attributes for the axes and so on. Ben Shneidermann has shown that only when there are less than 25 stars, their names, or other important attributes can be written, otherwise, it is impossible to read them due to overlaps. For deciding how to assign object attributes to starfield key-attributes, some rules can be applied:

- for the *x*-axis, ordinal variables are the best, time being one of them
- for the colour, nominal variables are the best
- for the *y*-axis, ordinal variables are also the best, but also nominal variables can be used.

For instance, suppose we have a database of persons such as

```
PERSON (Person-ID, last-name, first-name, middle-name,
date-of-birth, annual-income, profession, degrees, SEG,
etc.)
```

The attributes for the starfield display can be chosen as follows (Figure 7.13),

- age for the *x*-axis
- annual income for the *y*-axis
- colours to distinguish the socio-economic groups (SEG).

In an urban context, the database of building permits can be accessed using dates for the *x*-axis, floorspace for the *y*-axis, and construction categories with colours (see Figure 7.14)

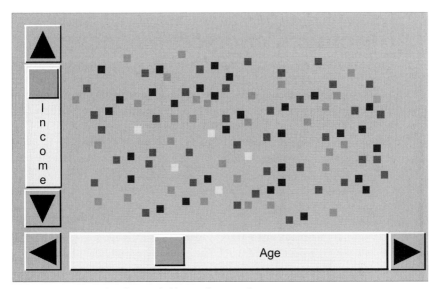

Figure 7.13 Example of a starfield visualisation for persons.

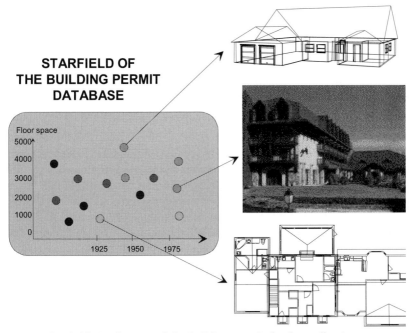

Figure 7.14 Starfield visualisation of the building permit database allowing access to various types of multimedia information; the circles act as anchors.

```
PERMITS (permit-ID, parcel-ID, permit-address,
applicant-name, applicant-address, floorspace,
date, draft-ID, dossier-ID, state, etc.)
```

The stars (here sometimes represented by squares or circles) can be seen as anchors to information about the permit they represent as exemplified in Figure 7.14 in which the starfield is issued as an entry mechanism (anchors) for multimedia information and possibly hypermaps.

To conclude this section on starfields, we can see them as a good illustration of the principle Overview, Zoom and Filter, Details on demands. At the opening, the overview is given by the starfield. Then, Zoom and Filters are essentially provided by the sliders allowing reduction of research space, and by the widgets to modify axis attributes. Finally when we have less than 25 stars, additional details can be displayed on demand, perhaps Permit-ID or applicant name.[4] For more details regarding starfields, see the FilmFinder project at the URL: http:/www.cs.umd.edu/hcil.

Treemaps

Treemaps (Johnson and Shneiderman, 1991), or more precisely space-filling treemaps are an elegant way to visualise objects organised hierarchically: they can be considered as a mix of two metaphors, the Venn diagram as exemplified in Figure 7.15, and the shelves for books. More exactly, when the shelves are vertical, and then horizontal and so on, to give a sort of nested treemap (Figure 7.16). Finally, the basic structure for database entry is given in Figure 7.17. Taking the concept *Overview, Zoom and Filter, Details on Demand*, we therefore have the Overview step. Then by zooming and filtering, we can reduce the search field of research. Finally, when the class of object of interest is found, we can jump to the starfield presentation system.

An example of a treemap is given Figure 7.18 for entering in the documentation relative to urban planning. In Figure 7.18, successively depicted are the initial overview (Figure 7.18a), then some high level details (Figure 7.18b), then zooming (7.18c), and zooming again (7.18d). Suppose that somebody was looking for new bye-laws concerning social care, now he can access to the desired documents, eventually organised as a starfield. To conclude this very quick introduction to global visualisation systems, let me

---

4   The positioning of labels such as permit number or applicant name is very similar to the problem of name placement in cartography, but with the additional constraint of real time. So this positioning has to be done in the fly, that is to say in real time during the display, and not before in several passes (crude positioning plus several refinement steps) as usually done in name placement in cartography (see Tanin *et al.* 1996) for details.

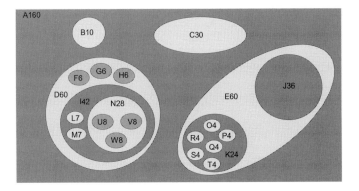

Figure 7.15 Venn diagram.
From Johnson and Shneiderman (1991).

say that starfields and treemaps can be considered as very powerful mechanisms for entry in urban databases.

7.3.2 Navigating in the information space

As visualising the global content of a database is something very interesting, navigating in the database is also important. By navigating, we mean going from one piece of information (perhaps a document) to another. Chaomei Chen has recently proposed a metaphor for navigating through documents

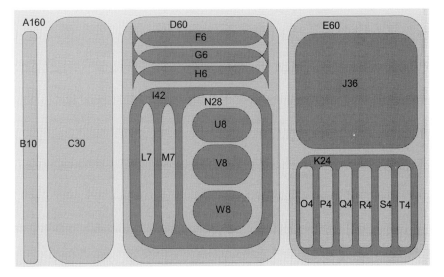

Figure 7.16 Nested treemap.
From Johnson and Shneiderman (1991).

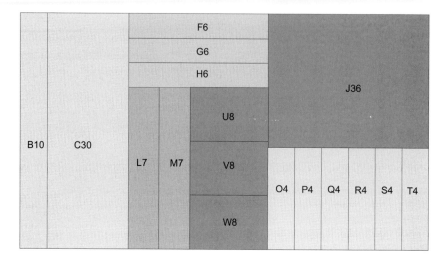

Figure 7.17 Space-filling treemap.
From Johnson and Shneiderman (1991).

(Chen C. 1998, 1999), possibly planning documents. The idea is to represent each document as a planet in this space, and to navigate as in a space vessel. Documents with similarities must be adjacent in this space.

Finally, in this space, the Chen model is as follows: each document is represented by a sphere and links to documents are symbolised by cylinders linking the spheres, the characteristics of which are given in Table 7.1. For details, please refer to http://www.brunel.ac.uk/~cssrccc2. An example of such a space forming a sort of 'spacescape' is given in Figure 7.19 first without queries, and then with queries (Figure 7.20).

Table 7.1 Chen visualisation model of key-elements

Visualised objects	Geometry	Visual attribute	Semantics
Document	Sphere	Radius	File size
		Colour	Year of publication
Link	Cylinder	Radius	Document-to-document similarity
		Length	Minimal semantic distance
Query hit	Cylinder	Height	Query-document similarity
		Colour	Distinct keyword

212

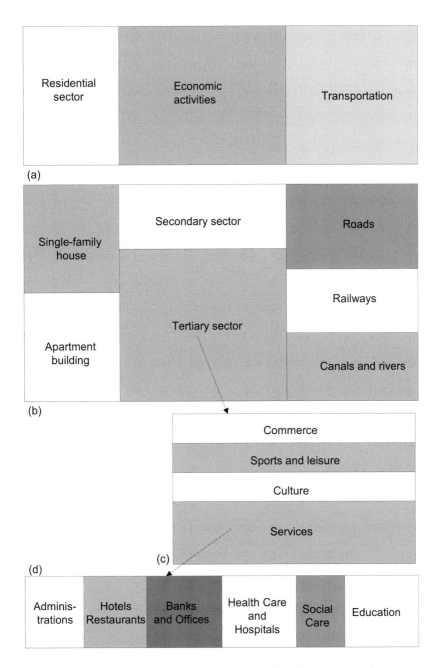

Figure 7.18 Example of a global visualisation system based on treemaps for accessing some documentation for urban planning. (a) Initial screen for the overview. (b) Screen after the first details. (c) Zoom on the tertiary sector. (d) Zoom on Services.

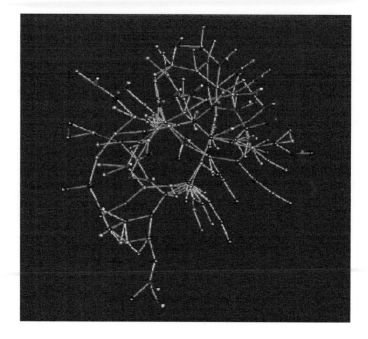

Figure 7.19 Overview of an example of the entire collection of documents (Chen 1998). Published with permission. Reprinted from *Journal of Visual Languages and Computing* 9 3, C. Chen 'Bridging the Gap: the use of Pathfinder networks in visual navigation' 267–86 © 1996 by permission of the publisher Academic Press.

7.3.3 The city as a metaphor for organising information (Dieberger and Tromp 1993, Dieberger and Frank 1998)

An interesting idea is to use the metaphor of a city for storing and accessing documents.[5] According to Dieberger and Frank 1998, cities are very complex spatial environments and people know to get information, infrastructure, etc. They also provide a rich set of navigational infrastructure that lends itself to creating sub-metaphors for navigational tools. So, a city metaphor makes this existing knowledge about a structured environment available to the user of a computerised information system.

The Information City is a conceptual spatial user interface for large information spaces. It is based on structures found in real cities, on knowledge of city planning and on how people learn such environments. It is an excellent source for a metaphor because it is extensible and can be navigated using commonly available infrastructures. In a real city people seldom really get lost,

---

5   Here the metaphor of a city is used to organise documents relative to urban planning; in other words, the city metaphor is used to access city information. Perhaps we can coin the expression of information city, or 'Meta-city'.

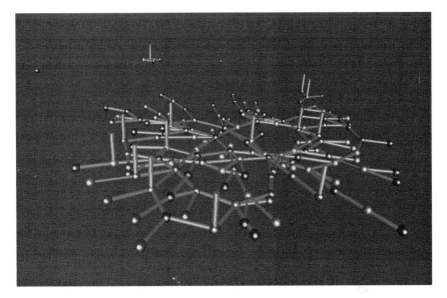

Figure 7.20 Landscape view with keyword search hit bars (Chen 1998).
Published with permission. Reprinted from *Journal of Visual Languages and Computing* 9 3, C. Chen 'Bridging the Gap: the use of Pathfinder networks in visual navigation' 267–86 © 1996 by permission of the publisher Academic Press.

first because there are always other people they can ask for help and second because a real city is an environment filled with various sorts of information that help in navigating. There are street signs, traffic signs, posters, landmarks, buildings that have a certain meaning for the user. Thus it seems intuitive to make use of the everyday navigation skills we use in real cities to navigate a complex computer-generated information landscape.

In the information city:

- **Buildings** are containers for information; they have a unique address and show their accessibility using doors; in other words, hypertext documents are visualised as houses in the information city.
- **Landmarks** are special non-access or public access buildings.
- **Rooms** are containers inside buildings; their walls may contain doors or windows to access other rooms or the outside.
- **Paths** connect two locations in the city. Paths outside buildings are visualised as streets or roads.
- **Intersections** of paths are squares. Large squares are major elements in the city.
- **Walking, driving and flying** are different ways of connecting one building to another for short- or long-distance navigation.

Continuing the metaphor, Figure 7.21 shows information within a building, first seen linearly, and then with a fisheye view.

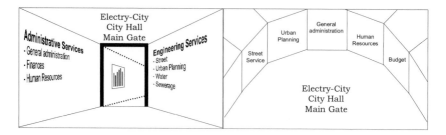

Figure 7.21 Two submetaphors in the information city.
According to Dieberger-Tromp, 1993). (a) The linear structure and the door metaphor leading to another information room. (b) The fisheye view. Source with modifications: http://www.mindspring.com/~juggle5/Writings/Publications/VRV.html

An example of the City Information metaphor applied to the city of Bologna, Italy is given Figure 7.22. For instance, clicking on the shop on the Virtual Bologna, you can get information about the commerce in the city, by clicking on the aircraft, information about the airport and airlines, and so on. For details, refer to (http://www.nettuno.it/bologna/mappaWelcome.html).

7.3.4 Example on organising information for flood history and prevention

For flood prevention it is important to know the history, often named the risk memory. Taking a river into account, over decades perhaps centuries, many documents have been registered from newspapers articles, photos, interviews, maps, videos and so on. For example, in Coulet and Givone (1995) and Coulet, Laurini and Givone (1995), the visualisation of the information related to flooding is made according to the starfield technique where the selected axes are milepoints along the river and time (Figure 7.23). In this example a typology of the documents was carried out in such a way that the red marks correspond to reports, blue marks to interviews, and green marks to photos, etc. The hydrographic network is schematised and one can add a few names of towns or villages as landmarks. This schema represents the overview step. One can, through zooming either on the hydrologic network or on the right inside window, delimit in a finer way the zone under study, the time range and the type of document. Finally, a document may be immediately visualised by clicking (details on demand).

So, documents are represented either by segments, that is to say, for a distance interval along the river and a date, or by rectangles when the flood has some duration over time. In addition, a semiology based on colours or on line styles can help to distinguish written reports, photographic documents and video documents. By clicking on a precise document, we can display it. And against this system, we can pose queries such as 'give me the documents describing the river Ofnoreturn, from mile 122 to mile 244 and from 1920 to 1990'.

Benvenuti a Bologna e dintorni...
Welcome to Bologna and its neighbourhood...
Mappa virtuale della città - Virtual map of the city

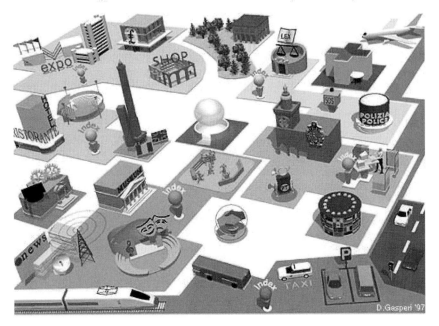

Figure 7.22 The virtual city of Bologna, Italy as an entry point in the web system. Source: http://www.nettuno.it/bologna/frame-mappa.com. Published with permission.

7.4 Conclusions: from static to dynamic, from passive to active

It is clear that conventional maps can be called 'static' because there is no dynamic updating or evolution: we can then consider now 'dynamic' maps. The other new aspect is the connection with other elements. The concept of hypermaps can be presented in a different manner, maps are not passive and they include multiple active links, giving 'active' mapping. By using animation, and especially real-time animation, new kinds of visualisation systems can be proposed to users and planners. In other words, cartography will pass from static and passive to dynamic and active.

As illustration, an example of the front page of a system for urban planning is depicted in Figure 7.24. It consists of three windows:

- the first one is for the retrieval and management of thematic information

217

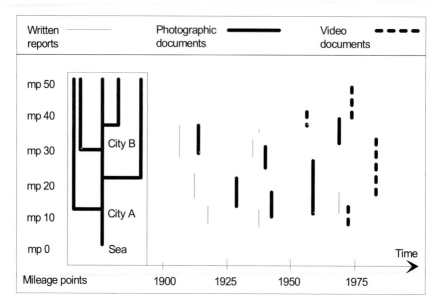

Figure 7.23 Example of entry display for flood history allowing access to documents of several media.

- the second is for spatial information, namely a small map of the city allowing to direct access to zones of interest
- and the last one for groupware activities, which is the subject of the next chapter.

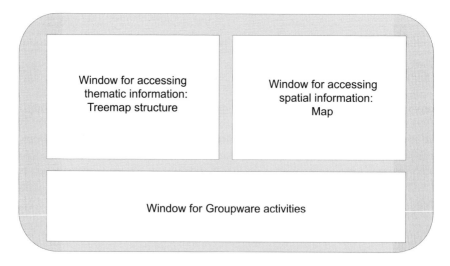

Figure 7.24 Front page of a system for urban planning.

8

GROUPWARE[1] IN URBAN PLANNING

In recent decades, computers have been mainly used for designing urban maps and storing appropriate data for urban management clerical works and to a lesser extent, for spatial analysis. But now, by means of the Internet, computer-supported co-operative works (CSCW) can be easily organised (Gronbaek *et al.* 1993, Malhing *et al.* 1995, Palmer *et al.* 1994). Known also as groupware these new techniques allow several persons, located in different places, to work together. This new mode of working appears very appealing for urban planners, especially for designing and assessing alternatives. One very important aspect is the use of GIS for public participation, especially facing new urban projects, but this issue will be developed in the next chapter.

The scope of this chapter will be to give the main elements of groupware and workflow in general, and in particular that dedicated to urban planning. After some definitions, we will examine carefully the consequences of the work of urban planners.

8.1   What is groupware?

Under different names such as groupware, CSCW and participatory design, new techniques are presently emerging allowing different people to work together for a very precise target with the assistance of computers. Indeed, the vast majority of existing computer tools are devoted to single users, or when dedicated to several users, the design of such co-operative systems is much more complex. Indeed, the main reason is because people supposedly working together are not located at the same place and not working at the same time, whereas clerical processes should be carried out within scheduled time.

---

1   A preliminary version of this chapter was published under the title 'Groupware for Urban Planning: An Introduction', *Computer Environment and Urban Systems*, vol. 22, 4 July 1988, pp. 317–33.

Urban planning is a task requiring several users to collaborate in order to design a plan including maps and written statements. Until now, computers were used mainly for cartographic purposes or as databases (Laurini and Thompson 1992) and it could be an important challenge to offer urban planners a CSCW system in order that the participatory design of urban plans can be performed (Laurini 1995). The goal of this chapter will be to define the potential of these new techniques for urban planning and especially for land-use planning. The first section will be devoted to the definitions of concepts of groupware and of connected techniques. Then we will try to summarise the benefits and limitations in order to conclude this section by presenting some elements in software architecture for co-operative information systems.

8.1.1   Definitions

According to Coleman[2] and Khanna (1995), 'Groupware is an umbrella term for the technologies that support person-to-person collaboration; groupware can be anything from email to electronic meeting systems to workflow'. But in reality, there are many other definitions of groupware. According to Nunamaker, Briggs and Mittleman (1995), groupware is defined as 'any technology specifically used to make groups more productive'. In Table 8.1 several technologies that fall under the groupware umbrella can be found. Besides supporting information access to several persons, groupware can radically change the dynamics of group interactions by improving communications, by structuring and focusing on problem-solving efforts, and by establishing and maintaining an alignment between personal and group goals. Schäl (1996) gives a similar definition.

According to Turban and Aronson (1998 p. 319), an example of the major benefits of video-teleconferencing are:

- providing opportunity for face-to-face communication for people in different locations, thus saving travel time and expenses
- enabling several members to communicate simultaneously
- providing the possibility of using several types of computer media to support conferencing
- enabling usage of voice (which is more natural than using keyboards).

More precisely, for our concern, four categories of grouping aimed at greater productivity can be distinguished:

1   group calendering and scheduling to automate the process of setting up the meeting and the collaboration

---

2   Be careful, there are two David Colemans working in this area, one in California, USA (David E. Coleman) and one in New Brunswick, Canada (David J. Coleman).

Table 8.1 Several definitions of groupware and related issues.
According to Nunamaker, Briggs and Mittleman (1995)

Groupware is ...

Computer Supported Co-operative Work (CSCW)
Group Decision Support System (GDSS)
Group Support Systems (GSS)
Co-ordination software
Group memory
Information filtering
Electronic conferencing
Groupware
Group scheduling
Team calendar
Group development tools
Team database
E-Mail
Project management
Group conferencing
Video teleconferencing
Electronic brainstorming
Shared drawing
Electronic meeting systems
Workflow automation
Electronic voting
Shared edition

---

2   electronic meeting support systems to increase meeting output, pro-
    ductivity, and the quality of decisions
3   group project management software for meeting follow-up
4   workflow software to route and track documents and action items
    generated from the meeting and other events.

As seen in the previous definitions between users, one of the main objectives
is not only to help people working together, but overall to provide them tools
in order to make them more productive, that is to say first of all that the
quality of communications between people should be enhanced. In order to
ameliorate the discussions, several sophisticated tools have been provided such
as computer-based conferencing, electronic meeting rooms, etc. Table 8.2
presents a general diagram emphasising the environment of Groupware under
the umbrella of Enterpriseware. In this diagram, on one hand there is an
information system (named database or repository, etc.) storing all
information concerning the enterprise and the group management system. In
the second part of the diagram, several layers are defined from the top
(enterpriseware) to the bottom (hardware), so groupware can be defined as an
intermediate layer.

Table 8.2 The Groupware environment
According to Coleman and Khanna (1995)

ENTERPRISEWARE		
	Cross-vendor support Integrated networks Executive information systems	Standards Local/remote servers
Database or infobase. Object repository or knowledge base. Document and image repository.	GROUPWARE Group Decision Support Systems Desktop video and audio conferencing Group application development environment Group editing Workflow	
	Group-enabled applications	E-mail messaging	Calendering Scheduling
	Personal productivity applications	Network operating systems	
	Operating systems		
	Hardware infrastructure: cables, multiplexers modems, ATM, frame relay, ISDN		

8.1.2 Participatory design

In several applications, the objective is that people should design something co-operatively. For instance, several kinds of engineers are involved in the design of cars, planes, bridges, buildings. In this case, not only communications must be enhanced, but also the software has to provide specific tools in order that at each step of the design, every involved member should do his work soundly. In order to reach this goal, a database storing several versions of the design, and also of the interactions between all engineers must be created. Here we have to address the problems of storing items of design at different steps and of different versions.

8.1.3 Benefits and limitations

According to Coleman and Khanna (1995), among the benefits are:

• increased productivity
• better customer service
• fewer meetings

- automating routine procedure
- integration of geographically disparate teams
- better co-ordination globally
- leveraging professional expertise.

But there exist some limitations to these techniques:

- there is too low a level of education in the business community about groupware
- organisations are resistant to change
- there are few standards in the groupware market.

8.1.4   Co-operative information systems

Under the name of co-operative information systems, one defines a database storing all information and knowledge necessary to support the collective work. Generally speaking, it consists of a distributed database system with one central database and several local databases. Whereas the local databases store information necessary for end users, the central database goal is to store common information for global project management. As exemplified in Figure 8.1, the distributed and co-operative information system is linked first to a system for task and message management with a second link for participatory design.

An important way to classify the different systems is the possibility for users to be at different places and/or at different times. Of course, when everybody meets at the same place, this is simpler, but in other cases, social interaction needs more complex computer-based systems. Figure 8.2, from Johansen *et al.* (1991) defines four possibilities according to the time and the place and gives several examples. Table 8.3 lists some classes of computer tools that enter into these options.

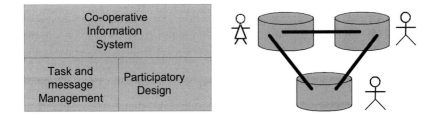

Figure 8.1 Structure of a co-operative information system.
Reprinted from *Computers, Environment and Urban Systems* **22** 4, R. Laurini 'Groupware for Urban Planning: An Introduction' 317–33 © 1998, with permission from Elsevier Science.

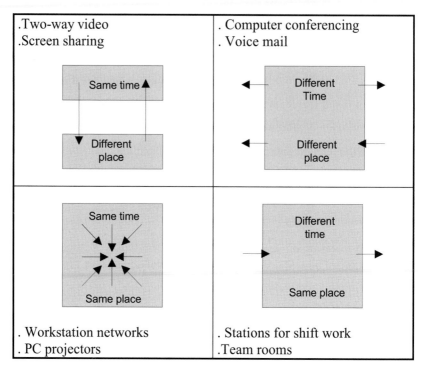

Figure 8.2 The 4-square map for groupware options.
From Johansen *et al.* (1991). Reprinted from *Computers, Environment and Urban Systems* **22** 4, R. Laurini 'Groupware for Urban Planning: An Introduction' 317–33 © 1998, with permission from Elsevier Science.

8.1.5   Group decision support system (GDSS)

Important characteristics of a GDSS may be summarised as follows (Turban and Aronson, p. 353):

- it is designed with the goal of supporting groups of decision-makers in their work; as a consequence the GDSS should improve the decision-making process or the outcomes of groups;
- it is easy to learn and to use. It accommodates users with varying levels of knowledge regarding computing and decision support;
- it is designed to encourage activities such as idea generation, conflict resolution, and freedom of expression.

An example of a room especially dedicated to group decision-making is given in Figure 8.3.

Table 8.3 Social interaction supporting systems
From Erik-Andriessen (1996)

OPTIONS	(a) For different place/different time	(b) For different place/same time	(c) For same place/same time
1. *Ad-hoc* information exchange (very short)	• Fax • Email general, or with information filtering and conversation structuring	• Telephone • Cellular phone • Videophone	
2. Group meeting/ presentation/ teaching (few hours)		• Video-conferencing systems • Audio-conferencing systems	• Group decision room (GDR) equipment • Presentation equipment
3. Teamwork (long time) (including 1, 2 and 4)	'Keepers' • Computer-conferencing • Co-editing system • Group-CAD systems • Other sharing systems 'Synchronisers' • Group-calendar • Shared project planning • Shared workflow system	• Video-conferencing systems • Audio-conferencing systems • Screensharing	• GDR equipment
4. Socialising		• Media spaces	

8.1.6   *Workflow modelling*

The modelling of the activities pertaining to a workflow can be built on different concepts (Schäl 1996). For instance, Kappel *et al.* (1998) have recently proposed a model based on the following concepts:

• Activity ordering. Activities which are part of a certain workflow have to be co-ordinated with respect to their execution order by means of an activity ordering policy.
• User selection. An agent which is able to perform a certain activity has to be selected for actually executing the activity by means of a user

Figure 8.3 Example of a room for group decision-making at Queens University
    Kingston, Ontario.
Source: http://www.groupsystems.com used with kind permission of Ventana
Corporation.

    selection policy; in some cases, an alternative agent can be selected by
    delegation.
- Worklist management. Each user is assigned a worklist possibly belonging
  to several activities.

The control flow between activities is specified by means of an activity network
relating activities to each other. They can be classified along several dimensions:

- control structures such as sequencing, branching and joining (Figure 8.4);
- dependency, the starting point of a successor activity may be defined as
  either the end of the predecessor activity, or the start thereof. So, two
  kinds of dependencies can be defined, either end-start (which is the
  default), or start-start;
- logical connectors between activities, such as AND, exclusive OR and
  inclusive OR
- temporal constraints specifying that one or several activities must be done
  in a certain maximum time, or before a given date.

An example of a workflow network modelling the activity ordering is given in
Figure 8.5.

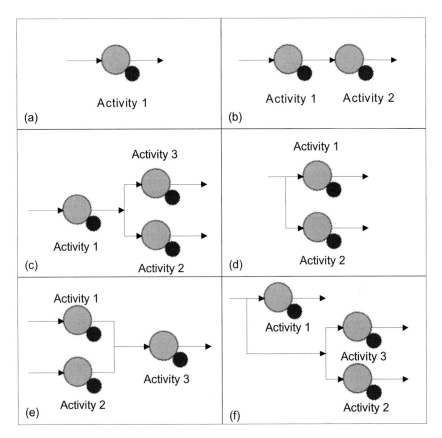

Figure 8.4 Workflow modelling. (a) an activity. (b) Sequence of activities.
(c) Branching process (fork). (d) Two activities in parallel. (e) and (f) More
complex ordering of activities.
From Kappel *et al.* (1998) with modifications.

8.1.7   Examples in geoprocessing

Two examples can be found in geoprocessing:

• the example (Coleman and Li 1999) for the management of geomatics
production, that is to say to ameliorate the relationships with the digital
map and chart providers. Figure 8.6 shows the main functionalities (Q/C
meaning quality control as described Chapter 4) and the interactions
between actors, Figure 8.7 depicts the status of the processing and Figure
8.8 some graphical report.

• one other example (Maack 1999) for street data maintenance, that is to
say to integrate and check updating data regarding streets. Several kinds
of actors can be distinguished, contributors, integrators and checking

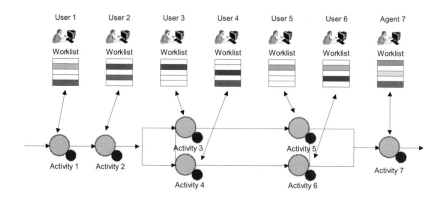

Figure 8.5 Example of a complex clerical process including several agents.
After Kappel *et al.* (1998).

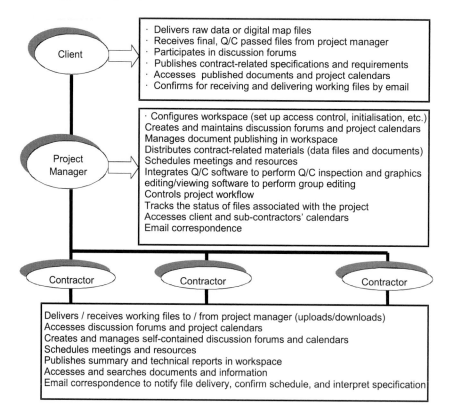

Figure 8.6 Functional data management and project management requirements of geomatics project participants.
According to Coleman and Li (1999), published with permission.

Index Map

Figure 8.7 An example of viewing file status.
According to Coleman and Li (1999), published with permission.

officer. Figure 8.9a illustrates the processes and the actors in the updating whereas Figure 8.9b gives the main actions and their sequence. First a contributor sends information to the integrator together with an attached document describing the corrections. Immediately the integrator e-mails an acknowledgement to the contributors and, if necessary, asks for verification. Afterwards, the database is updated and a final acknowledgement is sent to the contributor.

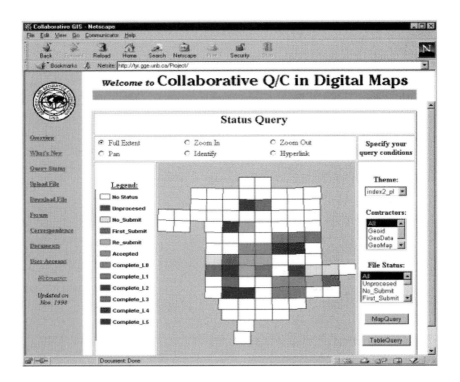

Figure 8.8 Example of database-driven production status reporting capability (Coleman and Li 1999), published with permission.

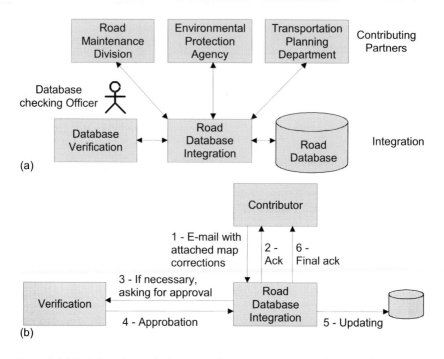

Figure 8.9 Workflow for updating street data. (a) Processes and actors. (b) Basic actions.

8.2   Groupware and urban planning

After having described some of the characteristics of groupware and CSCW, it is now important to answer the question of its relevancy for urban design and planning. Facing the importance of the task, the American National Center for Geographic Information and Analysis (NCGIA[3]) has created the initiative #17 on Collaborative Spatial Decision-Making, but with a broader sense than urban planning. For details, refer to the Scientific Report edited by Densham, Armstrong and Kemp (1995). Concerning urban planning, let us quote the works of Shiffer (1992, 1995) in the US and those of Maurer and Pews (1996) in Germany.

In this section, we will first describe an example of what could be done, the French planning design process, the actors and their roles, and the conditions of success for integrating groupware into urban planning practice in order to evaluate how groupware can be a relevant tool for enhancing those practices. We will finish with the presentation of some systems.

---

3   University of California at Santa Barbara.

8.2.1 Description of the French planning process

The French planning process is a very complex process involving several actors. In Figure 8.10, all the official steps of the planning process for a French city are mentioned together with the actors and the juridical actions, and some juridical deadlines. In fact in steps such as 'Blueprint design' and 'Possible Modifications', the staff of the Planning Agency and of other governmental offices are the main hidden actors whereas the Lord Mayor with the Local Council are the main visible actors. In other words, Figure 8.10 shows the routes that map and written documents must follow in order that the Plan will be approved and applied (Laurini 1980, 1982a, 1982b).

It is necessary to mention that this procedure can last from several months to a few years. In several cases, for some minor projects for instance, the renewal of a small city ward or the building of a bridge, the procedure is shorter even if the actors and the actions are more or less similar. Differently said, the plan evolves from version to version or alternatives. In order to get a

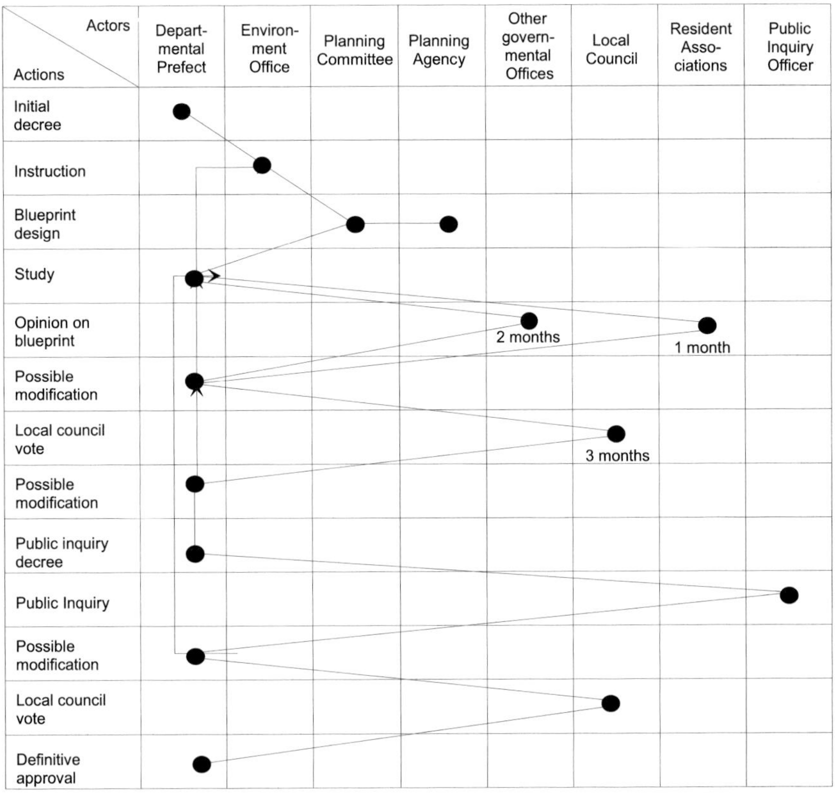

Actions \ Actors	Depart-mental Prefect	Environ-ment Office	Planning Committee	Planning Agency	Other govern-mental Offices	Local Council	Resident Asso-ciations	Public Inquiry Officer
Initial decree	●							
Instruction		●						
Blueprint design			●	●				
Study	● →							
Opinion on blueprint					● 2 months		● 1 month	
Possible modification	●							
Local council vote						● 3 months		
Possible modification	●							
Public inquiry decree	●							
Public Inquiry								●
Possible modification	●							
Local council vote						●		
Definitive approval	●							

Figure 8.10 The French juridical process of the French Land Use Plan design. According to Laurini (1982a).

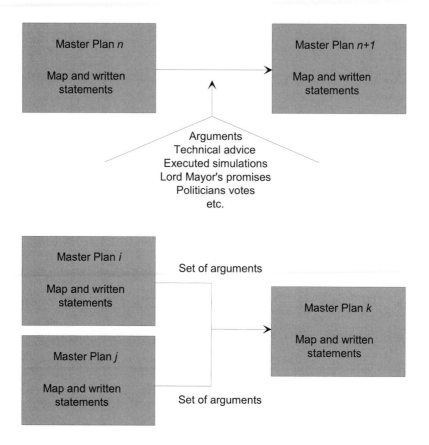

Figure 8.11 Information necessary to support new versions of plan.
Reprinted from *Computers, Environment and Urban Systems* **22** 4, R. Laurini
'Groupware for Urban Planning: An Introduction' 317–33 © 1998, with permission
from Elsevier Science.

new version, in addition to the differences in map and written statements, it is
necessary to store all information in order to know the reasons for this new
version (Figure 8.11). Some reasons may be technical such as arguments (see
Sections 6.6 and 9.4); technical advice and simulation results, and some others
may be more political such as citizens' opinions, elected politicians' votes or
vetoes, and so on. In some cases, a new version may not be the child of a single
previous version, but the 'argumented' amalgamation of several previous
versions.

8.2.2   Actors and roles in urban planning

In the planning process, we can consider that there exist three types of actors, politicians, technicians and citizens.

- By **politicians**, we mean in the narrow sense elected politicians in charge of urban planning, because they are the real decision makers regarding the future of their city, and in the wide sense, all members of the City Council who vote on decisions about the plan.
- By **technicians**, we mean all staff either working in Urban Planning Agencies or in the Engineering Services of municipalities or more generally of local authorities. They can be considered as advisers of elected politicians. They can use computers very widely in order to draw maps, to execute simulations, to write drafts of written statements and so on.
- In addition to isolated **citizens**, we must count the representatives of some urban groups (city dwellers' associations, etc.); those persons are generally asked to give their opinions at the beginning of the process, but overall they have to participate in the public consultation (public inquiry). Chapter 9 will be devoted to those aspects.

In general, groupware systems for urban planning must be addressed to technicians during their daily work and to elected politicians for key decisions and definitively during city council plenary meetings. More precisely, as shown in Table 8.4, depending on actors, a groupware system can have

Table 8.4 What groupware can afford to the main actors of urban planning

Actors in urban planning	Groupware in action	
	Frequency	Type of usage
Higher planning officer	From time to time, (minimum once a month)	General checking Final approval
City councillors in charge of urban planning	Several times a week	Requirements Meetings Simulation Votes
Other city councillors	Several times a month	Checking, votes Conferencing Meetings
City dwellers' associations	At the beginning and during all the processes, especially during inquiry	Collection of requests
Public consultation	At the end, daily, one month long	Photo-realistic visualisations Simulation Opinions
Urban planning staff and Municipal engineers and architects	Daily, during the whole process	Simulations, cartography meetings, authoring, messaging, conferencing

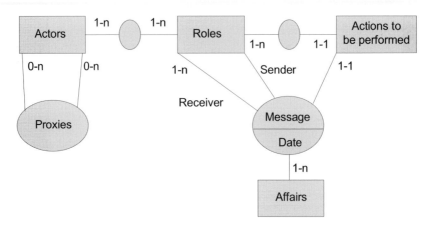

Figure 8.12 Relations between actors and tasks to be performed.
Reprinted from *Computers, Environment and Urban Systems* **22** 4, R. Laurini
'Groupware for Urban Planning: An Introduction' 317–33 © 1998, with permission
from Elsevier Science.

different utilisations. For instance, it will be used extensively all day long by
municipal engineers and architects, but only a few times by the Departmental
Prefect or the city councillors.

For all those actors daily working with such a system, especially urban
planners and municipal engineers, in order to follow the processing of the
plan, several messages must be sent overall regarding the actions to be
performed. An entity-relationship model of such information is depicted in
Figure 8.12, in which a message is defined as an association between affairs,
actions and roles. By roles, we mean the different functions that an actor must
perform during the planning process (possibly one for the majority of them).
By proxies, we mean the possible delegation an actor can give to another actor
especially when the former is absent, in order to speed up the whole process
overall for minor decisions or actions.

8.2.3   Conditions of success

After having very rapidly described the key elements of a groupware system
for urban planning, it appears important to examine the conditions of success
of such a tool; let us examine a few of them.

Will of participation

As a first key element, it seems that the will to organise urban planning work
with such a tool is very important, not just especially for all involved decision
makers, but also for technicians.

Training

One of the apparent drawbacks is that several urban planners are computer illiterate, so they can resist the change. In other words, training them in using computers will be of paramount importance.

Well-designed CSCW system infrastructures

Even if the CSCW product is well designed, it is important to have a reliable computing network and computers. In this category, we can also mention a very good quality urban database not only for census and alphanumerical data, but also for cartographic information. The coupling of a Geographic Information System and of a groupware system must be very efficient. According to Coleman and Khanna (1995) the equation of success for groupware is:

Groupware Success = Technology + Culture + Economics + Politics

As far as urban planning is concerned, personally, I think that the conditions of success are more or less the same as previously described.

8.2.4 Groupware in action

Suppose that a groupware system is installed within some local authority; the essential role will be the definition of the plan (map and written statements) for which different states can be defined (Figure 8.13). In order to fulfil this task, several versions of the plan must be considered. In other words, a graph

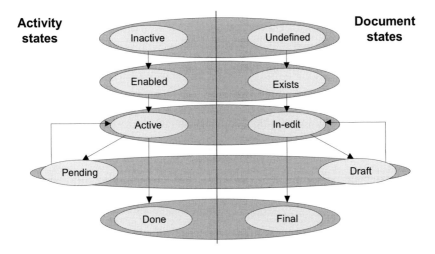

Figure 8.13 Coupling activity states with document states. According to Glance *et al.* (1996).

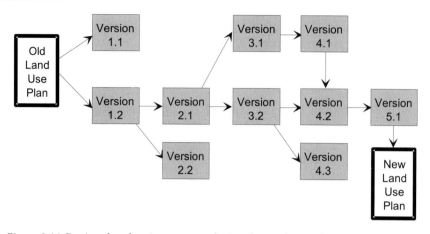

Figure 8.14 During the planning process, the Land Use Plan evolves through several versions before it reaches the final state, that is the approved new Land Use Plan.
Reprinted from *Computers, Environment and Urban Systems* 22 4, R. Laurini 'Groupware for Urban Planning: An Introduction' 317–33 © 1998, with permission from Elsevier Science.

of versions or alternatives can be defined starting from, for instance, the Old Master Plan to obtain the New Master Plan. In reality, two kinds of versions must be distinguished, proposals coming from the staff such as intermediate versions and versions duly approved by a vote of decision makers as exemplified in Figure 8.14.

When a modification of a version is suggested (ΔV), the concerned advisers will examine it both at technical and juridical levels (Figure 8.15). In some cases, financial consequences must also be evaluated especially for investment costs and returned taxes. If the suggestion is not accepted, perhaps some modifications can occur until a co-ordinator will either approve it or makes some arbitration. Finally, the suggestion is proposed to voters (elected politicians of the City Council) who can accept or refuse it.

Each zone can have a different way of processing routes: indeed, in some parts of the city, the agreement can be very easily reached, but in other precincts some difficulties or conflicts can occur. In other words, some states must be conferred to zones (Figure 8.16). But the geometry of those zones can vary over time. A sort of spatial automaton must be defined in order that each zone can follow the whole process until final approval and during the plan-making process, some zones can be split or some other zones can be amalgamated.

Sometimes a zone under study must be split into several smaller zones and different plans can be designed for those smaller zones. In Figure 8.17, a city was split into three zones, each of them being assigned several plans; the A zone has three different plans, the B zone two, and the C zone three. These

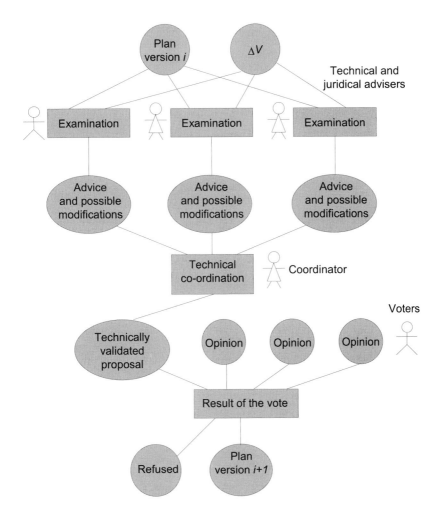

Figure 8.15 Each modification of the version (ΔV) needs to be examined by technical advisers in order to verify the technical and juridical aspects.
Reprinted from *Computers, Environment and Urban Systems* **22** 4, R. Laurini 'Groupware for Urban Planning: An Introduction' 317–33 © 1998, with permission from Elsevier Science.

different plans must be considered as different versions. Some of them are used to design new versions and some will no more be used to design new versions. When two versions of neighbouring zones are accepted they can be combined to give an amalgamated version.

By amalgamated versions, we do not mean only the carbon copy of the versions of the neighbouring zones, but also the necessity to fire some consistency tests at the boundary as exemplified in Figure 8.18. To get the

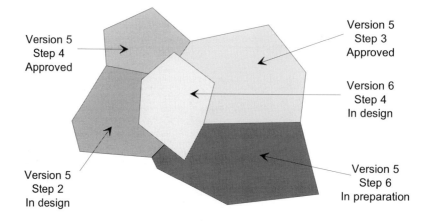

Figure 8.16 Each zone of the city can be at different states of approval. At the end of the process, for all zones, the planning proposals will be duly accepted.
Reprinted from *Computers, Environment and Urban Systems* **22** 4, R. Laurini 'Groupware for Urban Planning: An Introduction' 317–33 © 1998, with permission from Elsevier Science.

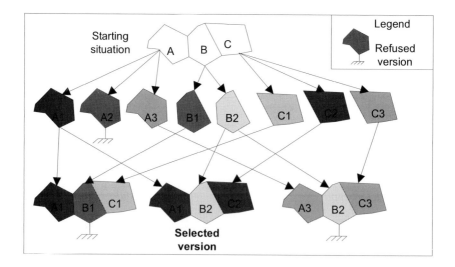

Figure 8.17 Graph of decomposition, and recomposition of versions.
Reprinted from *Computers, Environment and Urban Systems* **22** 4, R. Laurini 'Groupware for Urban Planning: An Introduction' 317–33 © 1998, with permission from Elsevier Science.

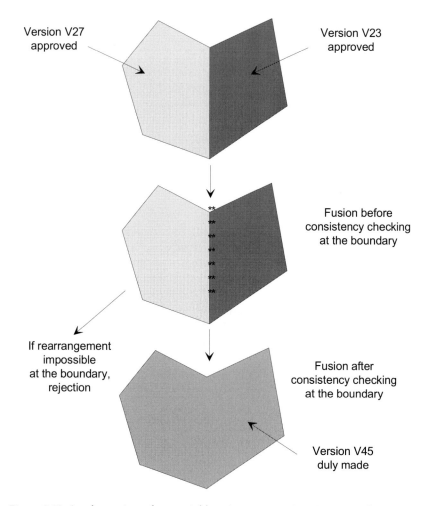

Version V27
approved

Version V23
approved

Fusion before
consistency checking
at the boundary

If rearrangement
impossible
at the boundary,
rejection

Fusion after
consistency checking
at the boundary

Version V45
duly made

Figure 8.18 Amalgamation of two neighbouring zone versions into a single zone
version, once consistency checking at the boundary is performed.
Reprinted from *Computers, Environment and Urban Systems* 22 4, R. Laurini
'Groupware for Urban Planning: An Introduction' 317–33 © 1998, with permission
from Elsevier Science.

final version all zonal versions will be amalgamated after several consistency
tests.

8.2.5 Towards systems for spatial negotiation

One of the main characteristics of a groupware system for urban planning will
be to offer actors the possibility to reach a certain level of consensus. For each
of them, all versions must be evaluated, not only on certain *a priori*-given

criteria, but also on the results of some simulation tools in order to estimate the possible long-term consequences. For example, a simulation tool can be used to estimate traffic noise from a traffic plan or to make a cost/benefit analysis for a Land Use plan. Whoever the actors and whatever their criteria, a spatial negotiation system must be offered to them. By spatial negotiation tools, we mean that some support will be offered to actors so that some level of agreements can be found. Such a system must take into consideration:

- the city and its current environment
- the version of plan and written statement under study
- the simulated consequences from different points of view
- the known actor's public criteria at global level together with their evaluation
- possibly some other aspects
- etc.

It seems important to mention that some (maybe all!) actors can have private criteria. Evaluations of those private criteria must be made at local level, i.e. on private or local computers, and the evaluation of such criteria must be made at local level. Anyway, an actor called the facilitator must propose some consensus to the concerned actors. Each of them must evaluate this proposal and give his/her agreement. When not all actors agree, perhaps the facilitator can impose a solution; a form of arbitration, the situation in which a consensus is imposed either by the facilitator or by the dominant actor.

8.2.6 *Example of collaborative GIS based on video-conferencing*

Recently Cowen *et al.* (1998) have presented a new collaborative GIS system based on video-conferencing especially devoted to industrial site selection in South Carolina.

> The industrial site selection problem appears to be well suited to GIS decision support environment. The selection of a site for a major investment is viewed as a multi-stage process in which initial criteria are specified for both regional and local factors. GIS provides an important way to filter sites on the basis of several criteria. Ultimately a number of prospective sites are identified and higher level management gets involved in the final decision. In fact, at this level the final selection can be based on personal or non-economic factors. In a competitive problem domain such as this perception can play a major role. It is important to give the decision maker the ability to gain a good representative view of the alternatives through a virtual navigation system that eliminates time-consuming drives to sites that do not meet expectations.

As said before, most of the collaborative decision-making solutions require that all the participants be located at the same venue. Very little attention has been given to ways that can efficiently expedite group decision making when participants are not physically at the same site. The assumption behind this approach is that there are executives and other high-level managers whose input is critical to the final decision. In many cases they may actually have veto power over the decision. It is also assumed that even the simplest GIS functions are still considered too complex for the average business executive. This will continue to be the case whenever someone is forced to leave their office and learn some new software. Therefore, it is worth a substantial investment to get even five minutes of their time to explore alternatives. In this environment it is critical that the system be responsive and all impediments to successful interaction be eliminated. The goal is to eliminate the interface and venue constraints by bringing the GIS and the expert to their site and providing both video and audio coaching. In a sense the **video conference** interface serves as chauffeur or instant access to a coach or 'expert'. The requirements for this system are:

- relatively inexpensive
- run on off-the-shelf personal computers
- uses commercially supported communication hardware and software
- supports unmodified commercial GIS software
- it is truly interactive in almost real time
- does not require GIS software to be installed on remote computers
- supports true collaborative sharing of the application
- requires almost no training
- allows participants to 'mark up' a common document and explore several alternatives.

The main components of the developed video-conferencing system are as follows:

1   A robust statewide GIS database that includes many layers of transportation, water supply systems, wastewater systems, landcover/land use, flood plains, demographics and economics, business and industry. For example this database includes every existing manufacturing plant, available industrial site, available building, water line and sewer line in the state.

2   Mature GIS Applications – InSite is a customised industrial site selection system developed by the South Carolina Department of Commerce. It utilises Visual Basic and Avenue to adapt ArcView

Figure 8.19 The video-conferencing system in South Carolina (Cowen *et al*. 1998). Published with permission.

into a high-level site-selection tool. It also includes an extensive set of digital images of the inside and outside of available sites and buildings. InSite is now utilised in the field by site selection professionals for site selection.

3   An interactive video-conferencing system as pictured in Figures 8.19 and 8.20.

4. A statewide Integrated Services Digital Network (ISDN) that supports point to point video-conferencing at 64 Kpbs.

8.2.7   GeoMed *(Schmidt-Beltz et al. 1998)*

The GeoMed project provides an Internet-based support for spatial planning and decision-making. It integrates support for co-operation, negotiation and mapping. A wide range of tasks of various user groups can be supported. GeoMed is a joint European project[4] with five developing project partners and

---

4   The project partners are: German National Research Centre for Information Technology (GMD), TNO-FEL and TNO-Bouw (The Netherlands), VUB (Belgium), Intecs Sistemi (Italy) and Intrasoft (Greece). The user partners are: the City of Bonn (Germany), the City of

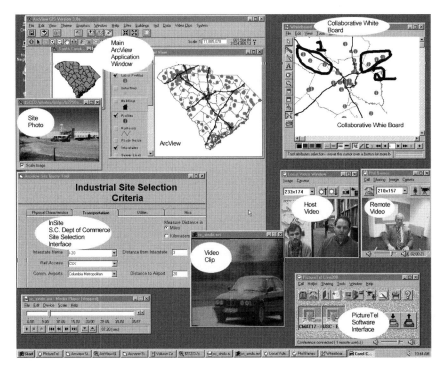

Figure 8.20 Example of the interface for industrial site selection based on video-conferencing based groupware system (Cowen *et al.* 1998). Published with permission.

four user partners. After a feasibility study in 1995, the project was launched in 1996. It consists of several components (Figure 8.21):

- basic support for co-operative work, especially by providing shared virtual workplaces
- GIS viewer which allows viewing of GIS data via the Internet, the data remaining on the server
- GIS broker and payment broker which allows new services where GIS data can be offered for sale
- discussion forum where users can discuss any topic. The discussion is structured as a tree or net of issue-position-argument hierarchies using the IBIS model (Section 6.6.2)
- software agents which perform notification and other services for users
- knowledge-based system application which allows analysis of plans with respect to special regulation and constraints on the basis of a knowledge-base.

Tilburg (The Netherlands), The Region of Tuscany (Italy) and the Technical Chamber of Commerce of Greece.

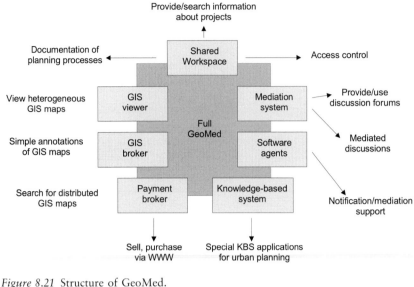

Figure 8.21 Structure of GeoMed.
From Schmidt-Beltz *et al.* (1998).

8.3 Conclusions: what groupware can afford to urban planning

As a conclusion, let us say that co-operative work and participatory design supported by computers are new techniques which can be applied to urban planning. Due to the specificities of the plan-making process involving several actors and dealing with zones of different states of processing, some particular techniques must be considered. In this chapter, emphasis was put on the necessity to consider such a system as a spatial negotiation system with version management. However, besides technical problems, difficulties remain in the acceptance of such a groupware system by all people acting in the planning process, especially by elected officials.

Due to the special architecture and also the fact that urban planners can have difficulties in specifying complete requirements of such a system, it is also important, before implementing such a tool, to discover all the characteristics, and to make some preliminary prototypes. Maybe other difficulties can occur when trying to connect some geographic information system to a groupware system in order to examine what the other technical difficulties are. Finally, I am confident enough to think that such tools will be in use in the next decade.

9

COMPUTER SYSTEMS FOR
PUBLIC PARTICIPATION

One of the key aspects of urban planning is public participation. The way citizens are involved in urban planning can vary a lot in different countries. Even if the number of existing experiments is very limited, I think it is of paramount importance in a modern democracy to offer city-dwellers some tools for, at maximum, designing their future urban environment or, at minimum, to be fully aware about it. In order to reach those ideals, among the problems to solve, let us mention

- **Participative plan design,** that is to say the way of involving citizens in the design of local plans; one example is the debate concerning environmental and urban planning.
- **Urban plan visualisation** is the way to present urban plans, not only map statements, but also written statements. Apparently map statements look easier to be understood, but studies have showed that a lot of people do not understand maps, especially when the contents bear some prescriptive juridical aspects. A direction of research can be the visualisation of urban plans in order to be understood by lay-people. Perhaps some combination of animated photos will help together with hypermaps systems (see Chapter 5).
- **Opinion collection** and synthesis: some people can give their opinions, or different remarks regarding the proposed plans (Laurini 1982a, 1982b). What kind of visual computer languages to offer them, especially in order to specify modifications of alternatives? What kinds of mechanism to provide for synthesising those opinions? The existing citizen forums provide an interesting solution to this issue.
- **Information distribution** and communication between citizens and the city council: for this task, the Internet can be used as a medium for exchanging information, ideas, maps between all actors.
- **Facilities organisations**: in order that scores of people with limited knowledge in computing can also participate by understanding plans, reading maps and written statements, giving their opinions and playing with urban virtual reality, the arrangement of the premises must be carefully studied.

In some cases, the expression 'Public Participation Geographic Information System' (PPGIS was used by several authors (Nyerges *et al*. 1997, or Jankowski 1998)) but, as we will see, we are long from a conventional GIS; and the expression 'Computer System for Public Participation' looks more appropriate in this context.

In this domain, we need to distinguish a very common character in the domain of public participation, named NIMBY (Not in my backyard!) representing people defending only their own property, often very aggresive in environmental disputes, and using general interest to protect their private interests. Sometimes, they 'pollute' the debate.

So the goal of this chapter will be to help planners in the design of new computer systems for involving citizens in the urban planning debate. This chapter is organised as follows. In the first section, the objectives of such a system will be clarified. Then the specifications will be launched, giving importance to debate modelling. And finally, some existing systems will be described.

9.1 Objectives for public participation and different ways of involving citizens

According to Craig (1998) organising public participation in a city can have the following objectives:

- **to expand the public's role** in defining questions and making decisions in which location or geography have a bearing on the issues addressed
- **to increase public participation** in the identification, creation, use and presentation of relevant information in various problem-solving contexts
- **to enable wider public involvement** of stakeholders in planning, dispute resolution and decision-making environments through a computer-based public participation process.

More practically, the public can be involved for the following collaborative planning processes such as public dispute resolution, facility siting/design review, futures and scenario planning. Different media can be used to involve people and an interesting classification was made by Vindasius in 1974, quoted by Sarjakoski, 1998 (see Table 9.1).

According to Schuler (1996), in order to be efficient, the characteristics of a public participation process should be:

- community-based, that is to say that everyone in the whole community/ city should be involved
- reciprocal, i.e. any potential 'consumer of information' should be a producer as well
- contribution-based, because forums are based on contributions of participants

Table 9.1 Types of public involvement, according to Vindasius (1974), quoted by Sarjakoski (1999)

Type of public involvement mechanism (Vindasius 1974)	Descriptive dimensions				
	Focus in scope	Focus in specificity	Degree of two-way communications	Level of public activity required	Agency staff time requirements
Informal local contacts	*	***	***	**	**
Mass media (newspapers, radio, TV)	***	*	*	*	*
Publications	***	**	*	*	**
Surveys, questionnaires	**	***	*	**	**
Workshop	*	***	***	***	***
Advisory committees	*	***	***	***	***
Public hearings	**	*	*	***	**
Public meetings	**	*	**	**	**
Public inquiry	***	*	*	**	**
Special task forces	*	***	***	***	***
Gaming simulation	*	***	***	***	***

Legend: * Low, ** Medium, *** High

- unrestricted, i.e. anyone can offer his participation
- accessible and inexpensive, that is to say that the use of the system must be free of charge to everyone
- modifiable, because of the legislative framework, the planning systems and the software can evolve, and these evolutions must be taken easily into account.

9.1.1 From Arnstein ladder to Kingston ladder

The problem of the various degrees of involving people in land-use planning is very old. In 1969, Arnstein proposed the first ladder for public participation with eight steps, manipulation, therapy, informing, consultation, placation, partnership, power delegation and citizen control. But this ladder was not seen as adequate. Starting from a previous work made by Weidemann-Femers some years before, Kingston (1998a) proposed a six-step ladder (Figure 9.1) which appears more relevant for our purpose. Among the steps, one can successively find from bottom to top (the lower steps meaning no real public participation):

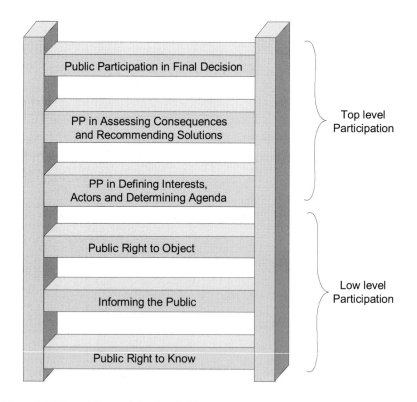

Figure 9.1 The public participation ladder.
According to Kingston (1998a, 1998b) with modifications.

- Public right to know: in this first-level phase, the public has only the possibility to be aware that some planning issue could be of interest.
- Informing the public: here the concerned local authority implements some action plan in order to inform the people but the people have no possibility to react.
- Public right to object: here the city-dwellers may say yes or no to a project, but have no possibility to react or to amend it.
- Public participation in defining interests, actors and determining agenda: this is the very first level of participation.
- Public participation in assessing consequences and recommending solutions: now the public is truly involved in analysing the impacts of possible decisions and can recommend solutions which can be accepted to be implemented.
- Public participation in final decision: this is real participation in the final decision; the decision is not only made by elected officers (city-councillors for instance), but each citizen can vote whether or not to accept the plan.

9.1.2 Nobre ruler

Another way to present the different scales of involving the public in the planning process is given Figure 9.2 (Nobre 1999). Nobre has established four main degrees of community participation: to inform, to consult, to discuss and to share. Lower levels are one-way procedures as they do not necessarily ask for any particular feedback from the community. On the contrary, higher levels of participation require two-way procedures as they imply capturing the public's reactions and feeding the decision-making process with such data.

The idea of planning secretly by distant professionals to avoid political or economic local constraints – a common concept from the 1960s – is nowadays completely overcome. In contrast 'to inform' (the first level Nobre is considering) is the minimal proceeding that one organisation must provide to assure any operation's success, whether a planning or a marketing operation.

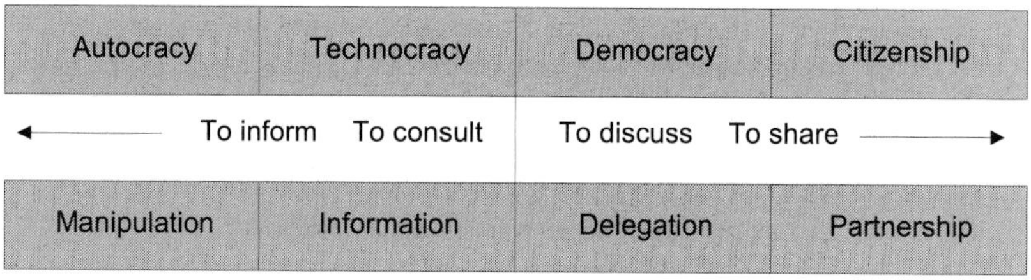

Autocracy	Technocracy	Democracy	Citizenship
← To inform To consult		To discuss To share →	
Manipulation	Information	Delegation	Partnership

Figure 9.2 Community participation ruler together with the political profile and the proceeding status (Nobre 1999).

The second participation degree is 'to consult'. It means not just 'to inform' but also to collect from some representatives' institutions their opinion, by organising public inquiries and discussion encounters. It can be considered a two-way procedure if and when the planning promoters accept the inquiries' results as an input in their decision-making process. 'To discuss' is somehow accepting 'to share' knowledge, but sharing in decisions is clearly the highest level of community participation. It is a turning point on this subject as well.

Being able to exert citizenship is as important as the will of the administrations to improve community participation in all urban life issues. Some theorists (Nobre 1999) speak about one 'educational city' to underline how urban fabric is a fertile field for innovative social behaviours. Yet, the usefulness of the 'participation ruler' would not be complete, without crossing it with the 'Proceeding Status' (the way information is provided and what it intends to achieve) and the 'Decisional Political Profile' (which kind of power exerted by what means (see Figure 9.2)).

9.1.3 Lessons learned in 22 years of public participation (Connor 1992)

Based on his experiences on more than 200 projects, here are the lessons learned by D.M. Connor. What follows is an attempt to summarise some key insights into the public participation process.[1]

- Understanding of the community is the essential foundation for an effective public participation process. In too many cases, organisations rush into print and start a 'dialogue of the deaf' because they have not taken the time to develop a systematic appreciation of the people and organisations in the community or region involved.
- The level and quality of participation by the public will be no better than that of the staff in the proponent's organisation. If public participation staff are managed in a top-down, traditional way, they are likely to manage the public in the same fashion. Alternatively, participatively managed staff are likely to work with the public more interactively since participation is part of the organisational culture. The development of a relevant public participation policy is often part of the preliminary work needed before launching a proactive program with the public.
- Spend no more that 20% of your resources trying directly to change the minds of the committed opponents of a valid proposal. Instead, direct your efforts to interest, inform and involve the usually silent majority and encourage them to deal with those who oppose the proposal.

---

1 In those experiences, apparently no computer tools were involved. My feeling is that even by using computers, those lessons still remain unchanged as a first approximation, and can be used as a basis for designing computer tools.

- Consensus is a noble ideal, but be prepared to settle for informed, visible, majority public support as a more realistic and achievable goal. Consensus decision making and alternative dispute resolution enjoy widespread interest today; both have a part to play but neither are panaceas. Consensus building is fundamental to most public participation programs, but when consensus is established as a norm for group decision making, everyone is given a veto. Single interest groups are likely to use it and perhaps abuse it.
- Social Impact Assessment needs a participative methodology and a focus on co-managing the impacts of a proposal, whether foreseen or not.
- In designing a public participation program, one secret of success is to try to ensure that everyone wins something, even if it is only recognition.
- Evaluation is the best way to learn from both your successes and your failures.

Considering those lessons, we can see that the fact of basing public participation on some computer tools will help a lot.

9.1.4   Vindasius classification

In 1974, Vindasius (quoted by Sarjakoski 1998) proposed a classification of the type of mechanism to involve people in planning (see Table 9.1). For each of these types, a sort of scale is given trying to evaluate the foci in scope, specificity, communications and so on.

9.1.5   Towards new tools for public participation

In Switerland (Brun 1999) there was some very narrow legislative framework for public participation with very formal procedures such as design of the project, publication, public consultation and decision. And the population was asked to give its opinion on a finalised project. The consequence was often deadlocks, especially with the NIMBY syndrome.

Due to several reasons, a negotiated involvement of citizens is possible, starting from the early stages of the land-use design. These processes imply a greater transparency from local authorities and public services; they need to provide complete and reliable information to citizens. As a consequence, new computer tools must be implemented to facilitate the debates and citizens' involvement as illustrated in Table 9.2.

9.2   First specifications of an information system for public participation

Now that we have clearer ideas regarding the objectives of a computer system for public participation in urban planning, it is possible to elaborate the specifications.

Table 9.2 Evolution of the environment for public participation (Brun 1999)

	Past	Present
Context	Growth Nationwide economies Institutional spaces Urbanisation	Cycles growth/recession Globalisation Areas with varied functions Metropolisation
Priorities	Controlled land use	Sustainability
Implementation	Quantitative and normative approaches Sectorial management Taylorism	Qualitative approaches Global management of complex realities Task integration
Participation	Institutional participation Formal procedures Restricted access to information	Negotiated involvement Interactivity Transparency
Information tools	Alphanumeric databases Drawing mock-ups Photo-camera	GIS–CAD Connected databases Multimedia
Information products	Maps Mock-up Photos Text files Calculation files	Raster and vector maps Aerial photos Satellite images Multimedia integration Simulation
Communication assistance	Paper Photos – slides Video	Data servers Internet – Intranet CD ROM

9.2.1 Roles and actors

According to Nijkamp and Scholten (1991), if the type of role for technicians is more or less easy to define in contrast to interested citizens, it is not so easy (Table 9.3) because their requirements are not very easy to know. Indeed, for some of them, their objective is to reach their goal, whereas for others, the means look more important than the goals.

Regarding computer functional architecture, recently Sarjakoski (1998) presented a general framework (Figure 9.3). In essence, this is a support-decision system in which there is a module for public participation and another for political evaluation. Among the peculiarities, let us mention the importance of images, (i.e. not only aerial photos but also any kind of photos, for instance, buildings, and also computer image generated or virtual reality as exemplified in Section 7.2.1.

For some other people (Nyerges *et al.* 1997), PPGIS (for Public Participation GIS) are new tools that foreshadow something more important, a new type of computer collaborative system the main characteristics of which are the following:

Table 9.3 Types of user demand for a GIS (Nijkamp and Scholten 1991, p. 17), quoted by Vico and Ottanà (1998)

Type of role	Information demand	User demand	Type of GIS
Information specialists	Raw data	Analysis Flexibility	Large Flexible
Preparers of policy	Raw data and pre-treated data (= information)	Analysis Good flexibility	Compact Manageable
Policy decision makers	Strategic information	Good accessibility to users; weighting and optimisation models	'Small and beautiful'
Interested citizens	Information	Good accessibility to users	'Small and beautiful'

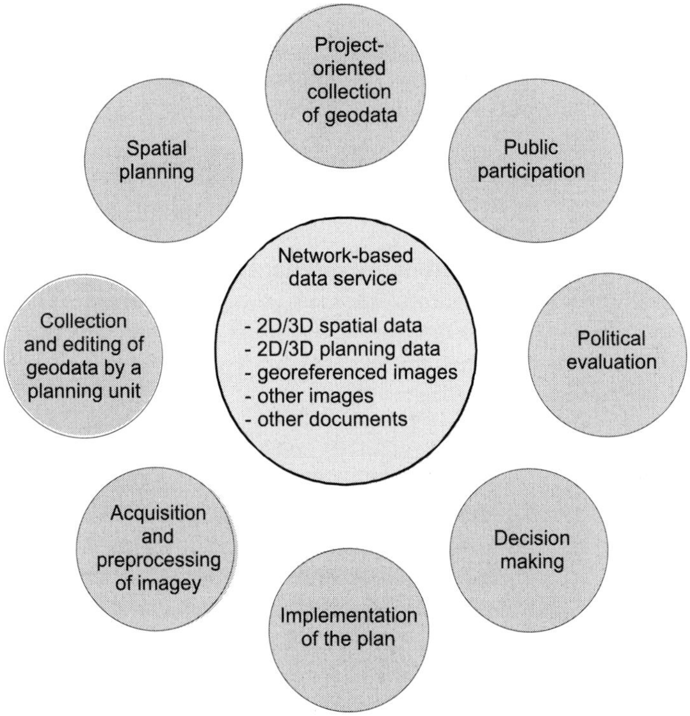

Figure 9.3 Outline of some of the processes related to spatial planning, including public participation (Sarjakoski 1998).

Table 9.4 Functional capabilities for PPGIS (adapted from Nyerges *et al.* 1997)

Level 1: Exploration/communication support	Level 2: Enhanced analysis/deliberation support
(1) **Group communication**: idea generation and collection through anonymous input, exchange and synthesis, identification of common ideas. *Tools*: data/voice transmission, electronic voting, electronic white boards, discussion groups, computer conferencing, and public computer screens.	(5) **Process models**: descriptive/ simulative models of physical and human spatial processes. *Tools*: GIS-embedded models, specialised models linked to GIS visualisation tools, intelligent agents, expert systems, knowledge bases.
(2) **Information Management**: storage, retrieval, and organisation of data. *Tools*: spatial and attribute database management systems.	(6) **Advanced spatial visualisation**: virtual realities, multimedia animations. *Tools*: see Chapter 7.
(3) **Graphic Display**: spatial and attribute data visualisation. *Tools*: shared and individual computer displays of maps, charts, tables, images, and diagrams.	(7) **Decision Models**: various decision rules integrating individual and group-derived evaluation criteria with alternatives performance data. *Tools*: Multi-criteria decision support techniques.
(4) **Spatial Analysis**: basic analytical functions *Tools*: proximity, buffering, overlay, data analysis, data mining.	8) **Structured group process**: facilitated/ structured group interaction, brainstorming. *Tools*: automated Delphi, nominal group technique, electronic brainstorming.

- role of participants – innovation/creation
- diversity of views, managing contradictions inconsistencies
- output dedicated to public
- e-mail, archives, real-time analysis
- handling plan history and alternatives.

So, two levels of systems can be defined. The first level can be defined as a support for exploration and communication between the actors, and more precisely with the citizens, whereas the second level should be more dedicated for enhancing analysis and deliberation between actors. For details see Table 9.4 where in addition some examples of computer tools are listed.

9.2.2 Towards new visualisation systems

Among new visualisation systems, there is virtual reality which deserves a very important section (Section 9.3) in this chapter. But the cartography of citizens' opinions is also something important. One key idea is to use some hypermap techniques to organise those opinions. According to Shiffer (1999),[2] we need:

- to recollect the past, by using some annotation mechanism for regrouping what was said, what was done, or what a place was like, etc. However, the lack of documentation or data to support this can lead to inconsistent individual memories. In addition, resulting arguments can dominate a discussion and shift the focus of a meeting from the matters at hand.
- to describe the present with some navigational aids; it is necessary to familiarise participants with an issue or area being discussed so that everyone can work from a common base of knowledge. The juxtaposition of media (maps, photos, thematic data, etc.) can strengthen a collective understanding of the various characteristics of a given site or issue. But the lack of access and (more recently) filtering for this information can handicap the description.
- to speculate about the future by using some representational aids (for instance virtual reality) and the extrapolation of measurable phenomena from past experience and application to the future using informal mental models perhaps more formalised using computer-based analysis tools. But the existing mathematical analysis tools are traditionally limited by speed difficulties and abstract output.

So, storing annotations can be done geographically, chronologically, by association, by relevance, especially by using 'post-it notes' (see Section 6.6.3). Of course, annotation types can vary from:

- simple graphical marks (such as lines, circles, dots, etc.)
- video sketching (graphical 'what if')
- textual annotation (flexible, low storage/bandwidth)
- audio annotation (fast, can be awkward)
- video annotation (expressive, compelling, storage/representation concerns)
- building blocks or mockups (simple, need more connection to digital representation).

9.3 Virtual reality for public participation

As presented in Section 7.2.1, several kinds of visualisation systems can be used for public participation, especially based on virtual reality. Taking again the Verbree and Verzijl (1998) and the Verbree et al. (1999) presentations, we can think of:

- workbench systems
- cave systems.

2 See http://yerkes.mit.edu/shiffer/MMGIS/Title.html for details and animation.

Both systems assume that a complete 3D model of the city already exists, that a specially equipped room also exists, and that the citizens are provided with some head-mounted displays or special glasses. More precisely, two models must exist, the present city, and the planned city.

For urban planning, an ideal virtual reality system can give the citizen the impression that he is present both in the actual and the planned environment. First of all it is necessary to fill the space as much as possible with the realistic representation of a model of the study area. With works developing the infrastructure this is often a combination between the existing reality and the new situation. This image has to be created from the model at the same time, which corresponds with the change of viewpoint (real-time rendering). This requirement sets up conditions for the hardware and software to be used as well as for the modelling itself.

The most affordable system is the screen of the PC as a 'window' to virtual reality. The user himself is not present in the system, but it is possible to present an image of the first-person on the screen. By offering nearly simultaneously an image for the left and the right eye through shutter glasses, the brain is capable of reconstructing a 3D-image. This 'Window-on-the-World' can be replaced by a projection on a screen from underneath a table, as on the Virtual Workbench (Figure 9.4) or by a projection on a large cylindrical screen (Theatre VR). The user of the system is still not really present in the projected world, but because of the large viewpoint the view becomes much better. In order to obtain a good stereoscopic representation a refresh rate of 2 × 30 per second is necessary. This is possible only with specially designed graphical hardware. Different from the other earlier presented forms of virtual reality, the cave gives users the opportunity to be actually present in a virtual world. This world can be typically created in a space of around 3 × 3 × 3 metres, in which on three walls (in front, on the left and on the right) and on the floor a multiple projection takes place. Also in this case use is made of 'shutter glasses' to evoke a 3D-image. Similar to the already mentioned systems several spectators have the possibility to be present. A part of the real world stays visible and manageable.

Figure 9.4 Virtual Workbench http://www-graphics.stanford.edu/projects/RWB/.

Figure 9.5 Virtual LA produced by William Jepson and colleagues in the Urban
    Simulation Team.
Source: http://www.aud.ucla.edu/~bill/UST.html. Image courtesy of William Jepson,
Director UCLA Urban Simulation Laboratory.

For an interesting panorama, please refer also to Dodge *et al.* (1998). One
other possibility is still to use non-immersive virtual reality technique, as given
for instance in Figure 9.5 for Los Angeles, California.

As an example, let me very quickly present the CommunityWorks software
(Figure 9.6). This product is a place-based decision support system for
community planning and design decision making. Geographic Information
System (GIS), 3D visualisation and simulation technologies have been
integrated in a system designed to be customised by each community.
CommunityWorks provides an interactive, real-time multi-dimensional
environment in which citizens and professionals can reach consensus on
goals, objectives/policies, and design the future of their community. Citizens,
planners, designers, and public officials will operate in a virtual world in real
time, and will have the ability to propose policies and formulate and design
alternative scenarios. Over time, they can see how these changes impact their
environment physically, fiscally and socially. For a similar example of another
company, see also http://www.multigen-paradigm.com/gallery.htm.

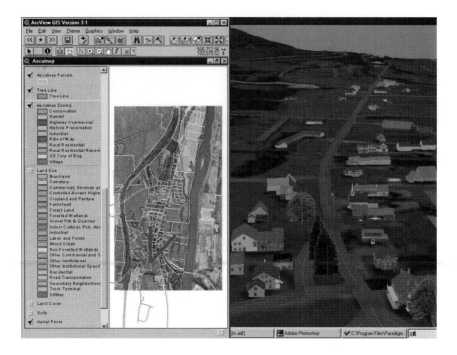

Figure 9.6 Example of project visualisation from CommunityWorks.
Source: http://www.simcenter.org/Projects/CPSP/CommunityWorks/
communityworks.html. Published with permission.

9.4   Examples of information systems for public participation

To conclude this chapter on information systems for public participation, let
us present three examples, the first one in the Twin Cities (Minnesota, USA),
the second in Idaho based on a Spatial Understanding and Decision Support
System, and the last one in the UK.

9.4.1   Public participation in the Twin Cities (Craig 1998)

In Minnesota, USA, both Minneapolis and St Paul (named the Twin Cities)
require input from citizens on any planning process. For instance, St Paul has
divided itself into 17 districts for citizen participation, each of them averaging
16,000 people. Any planning activity must go through a district council before
it can be taken up by the city. But Minneapolis has initiated its Neighborhood
Revitalization Program, the goals of which are to reorganise the building
neighbourhood, to create a sense a community and to increase collaboration.
This process involves six steps:

1   Develop a participation agreement that spells out how to proceed.
2   Build a diverse citizen participation effort. A neighbourhood revitalisation programme steering committee reaches out to the community to learn about issues, needs, and opportunities.
3   Draft a plan. This should address top issues in the neighbourhood with clearly defined objectives.
4   Review and approve the plan at the neighbourhood level.
5   Submit the plan to government for review, approval, and funding.
6   Implement the plan, that is to say 'The neighborhood organisation staff and resident volunteers help carry out, monitor and revise the plan as it is implemented. Cooperation with government staff, nonprofit organisations and the private sector ensures successful and timely implementation of the Neighborhood Action Plan' (Craig 1998).

In this process, one of the key issues was the creation of a web site (http://www.freenet.msp.mn.us/org/dmna/ including the following characteristics:

• hot topics (e.g., major change in membership rules, or copy of new Minneapolis Plan for comment)
• official documents (e.g., bylaws, copies of all correspondence)
• board meetings (director names with email links, meeting minutes)
• Neighborhood Revitalization Program details (e.g., official agreement, survey results, meeting minutes)
• information about the neighbourhood (e.g., Census data, address and other details for all residential buildings, political representatives with email links, skyway map and hours, business directory)
• links to local media stories (e.g., construction noise violations, downtown as a place to live, skyway system)
• links to related local sites (e.g., bus schedule, local government sites and publications, activity guides, Greater Minneapolis Convention and Visitors Association)
• links to national sites (e.g., International Downtown Association, Project for Public Spaces, National League of Cities).

9.4.2   Overview of SUDSS capabilities (Jankowski 1998)

Nyerges *et al.* (1997) analysed various scenarios of public decision problems and suggested that GIS-enabled public participation process involves three phases:

1   exploration of data to clarify issues
2   establishing a set of decision objectives
3   evaluation of feasible options of land-use planning.

These phases can be supported with tools classified at two levels of functional capabilities (see Table 9.3). Both levels can be treated as building blocks of

259

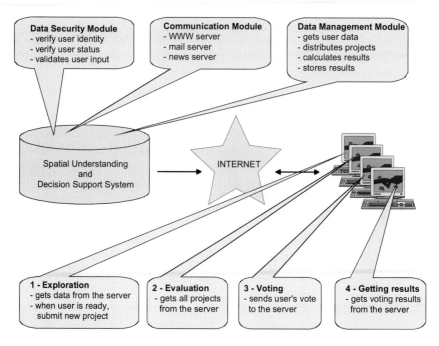

Figure 9.7 Server-client architecture of SUDSS for distributed public participation. From Jankowski 1998. Published with permission.

PPGIS in the sense that level 2 would not work without level 1, but level 1 could stand alone.

The goal of developing the Spatial Understanding and Decision Support System (SUDSS) was to allow participation in the decision-making process from different locations and at different times (distributed space and time). Several strategies for a design of SUDSS were considered. All strategies relied on the integration of GIS and decision support tools with the Internet infrastructure (see Figure 9.7).

The software is based on server-client architecture (see Chapter 10). The server component, working under the Windows NT operating system, assumes the database management tasks. The client (user) component is a stand-alone application which communicates with the database on the server. The data exchange between SUDSS client software and the SUDSS server is solely based on the TCP/IP protocol. See Figures 9.8 and 9.9 for copies of some user interfaces.

9.4.3   Virtual Slaithwaite (from Kingston 1998b)

At the University of Leeds, UK, an interesting planning exercise was performed in liaison with the village of Slaithwaite.

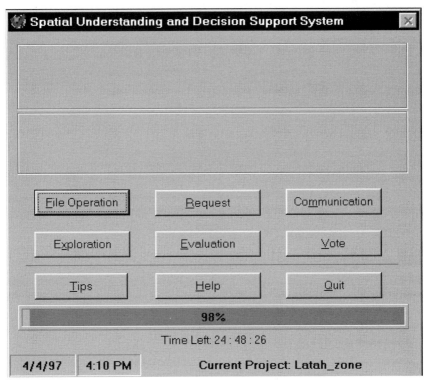

Figure 9.8 Main user interface window of SUDSS with tool buttons.
From Jankowski 1998. Published with permission.

Consultations with Kirklees Metropolitan Council (KMC) Environment Unit revealed that a Planning for Real (PFR) exercise was being
planned for early June 1998 in the village of Slaithwaite by Colne
Valley Trust (CVT), a local community action group. PFR is an idea
developed as a means of getting local people more closely involved in
local planning decisions through active participation and interaction
with large-scale models of the area in question. The Slaithwaite PFR
exercise was co-ordinated for the CVT by planning consultants and
part funded by the local council. A 1:1000 scale three-dimensional
model of a 2km^2 area of the Slaithwaite village and valley was
constructed by the CVT and planning consultants with the help of
local schoolchildren. This was used as a focus for local discussion
about planning issues within Slaithwaite. Local people were invited to
register their views about particular issues by placing flags with
written comments onto appropriate locations on the model. The
results of this exercise were then collated by the consultants NIF, and

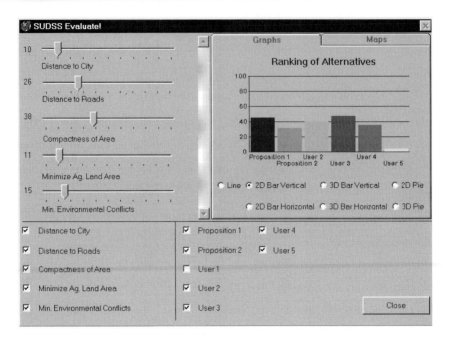

Figure 9.9 Working in the evaluation window the user can select evaluation criteria (lower left), assign criterion weights to the selected evaluation criteria (upper right), select alternative land-use plans for evaluation, (lower right), and view the evaluation results (upper right). The names alternative and proposition are synonymous and represent land-use plans.
From Jankowski 1998. Published with permission.

then subsequently fed back into the planning process through appropriate policy documents and plan formulation mechanisms.

The Slaithwaite PFR exercise provided this research project with an ideal opportunity to develop, pilot and live test a simple VDME that mirrored the functionality of the physical PFR model. This was called 'Virtual Slaithwaite' and can still be used on the WWW at: http://www.ccg.leeds.ac.uk/slaithwaite/. The virtual version of the exercise was launched on the web alongside the physical PFR model at a local village event organised and run by CVT. Six networks were available throughout the event for public use.

The design of the system revolves around a Java map application that allows the user to perform simple spatial query-and-attribute input operations (see Figure 9.10). Using this Java map applet, users can view a map of Slaithwaite, perform zoom and pan operations to assist in visualisation and navigation, ask such questions as 'What is this building?' and 'What is this road?' (spatial query) and then make suggestions about specific features identified from the map (attribute

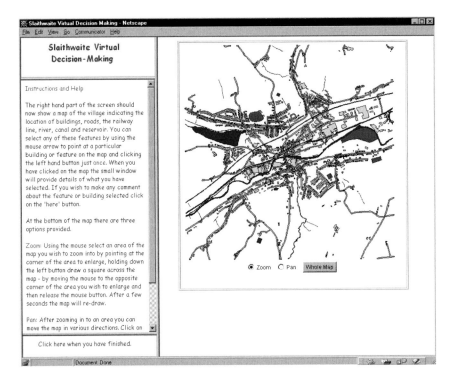

Figure 9.10 Example of virtual decision making in urban planning as a case study at the University of Leeds (Kingston 1998b) http://www.ccg.leeds.ac.uk/slaithwaite/).

Reprinted from *Computers, Environment and Urban Systems* **24** 2, R. Kingston, S. Carver, A. Evans, I. Turton 'Web-Based Public Participation Geographical Information Systems: An Aid to Local Decision-Making' 109–25 © 2000, with permission from Elsevier Science.

input). All user input is stored in the web access logs for future analysis and feedback into the planning process. In this manner a community database can be created, representing the range of views and feeling about planning issues in the locale. The 'Virtual Slaithwaite' web page is still available on-line after the PFR event and is still gathering responses from local people, as well as generating interest from further afield.

9.4.4 Decision support via the World Wide Web (from Kingston 1998b)

By providing access to appropriate data, spatial planning models and GIS via user-friendly web browsers the WWW has the potential to develop into a flexible medium for enhanced public involvement in the planning process. Several web-based systems can now be found on-line but the majority of these tend to be demonstration systems using sample data which are not necessarily problem specific and are therefore of little interest to the majority of the public. Many of these systems merely provide information in a uni-directional form such as listing planning applications and publishing development plans. In the UK, Devon County Council's Structure Plan was put on-line providing access to documents outlining the Council's strategic policies and proposals. Details on how to object to the proposals and the times, dates and places of meetings were also provided. But the system lacked any ability for the public to interact with the plan by populating the system with their own information, ideas or objections. Visit Devon County Council's Structure Plan web pages for details: http://www.devon-cc.gov.uk/structur/.

9.5 Conclusions

As described in this chapter, modern technologies permit radical change to the nature of public participation in decisions regarding urban planning. Only a few examples were given giving an hint of what will be possible in the future. Exchanging experiences between countries will be very fruitful, for instance under the aegis of associations such as the International Association for Public Participation (IAP2), whose goals (www.iap2.com) are as follows:

- Serve the learning needs of members through events, publications and communication technology.
- Advocate public participation throughout the world.
- Promote a results-oriented research agenda and use research to support educational and advocacy goals.
- Provide technical assistance to improve public participation.

For the future, some are forecasting the apparition of a new kind of citizen, named cyber-citizens, or sometimes cyber-spatial citizens, who will be citizens using new information technologies to act as real citizens, especially in connection with authorities.

To get more information about this increasingly important practice, please contact also the forum *ppgisforum@spatial.maine.edu*. See also the proceedings of the International Conference on Public Participation and Information Technologies, Lisbon, 20–22 October 1999 and their web site http://www.citidep.pt/icppit99/ for more details.

10

COMPUTER ARCHITECTURES
FOR URBAN PLANNING

The purpose of this chapter is to give the reader some elements in order to understand several aspects in hardware, software and network architecture for the applications dealing with urban planning. Indeed, the evolution of computers, and especially in the domain of telecommunications have allowed the possibility of novel architectures which are interesting for urban planning, and the more important appear to be the co-operative aspects. Successively, detailed information will be given for different computer architectures such as:

- client-server
- federation of several information systems
- interoperability
- architecture for groupware
- datawarehousing.

Aspects concerning real-time systems for environmental monitoring will feature in the next chapter.

10.1   Generalities about information system architecture[1]

At computer system level, several kinds of architectures for information systems are possible, centralised, decentralised, client-server, co-operative, etc. The last possibilities need to define some standards in order to make several systems interoperate. Let us examine these architectures.

10.1.1   Centralised databases

In general, traditional applications were implemented in isolated computers, possibly with several terminals connected to the computer and within an

---

1   Some elements of this section are taken from R. Laurini 'Spatial multi-database topological continuity and indexing: a step towards seamless GIS data interoperability' *International Journal of Geographical Information Science*, Volume 12, 4, June 1998, pp. 373–402.

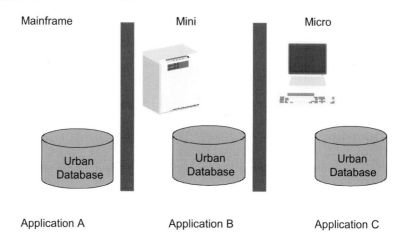

Figure 10.1 Separation between several urban databases

institution or a company, several computers existed with no connections between them. In other words, centralisation means the separation of different applications, each, or a small set, working on a single computer as shown in Figure 10.1. A first historical evolution (in the 1980s) was known as downsizing in which, instead of working with huge computers, several isolated dedicated micro-computers were used. Another evolution was to export terminals from the site itself to a remote location. In this case, one can speak indifferently about either remote terminals accessing databases or terminals accessing remote databases.

10.1.2 Multidatabase systems

The next evolution was to try to connect those independent computers (possibly mainframes, mini-computers or micro-computers) by means of a network under a variety of distribution architectures which will be explained and detailed later. With the general name of multidatabases, several functionalities of connected databases systems have been invented.

The great advantage of connecting or federating several databases together, are that users can develop or run tasks on several machines as they used to do on single machines. However, the big problem is technical since computers, operating systems, Database Management Systems (DBMS) and GIS can be very different in their structures so that making them communicate is a very difficult and heavy task. This section will present the problems and the possible solutions in order to federate several geographic databases and make them co-operate.

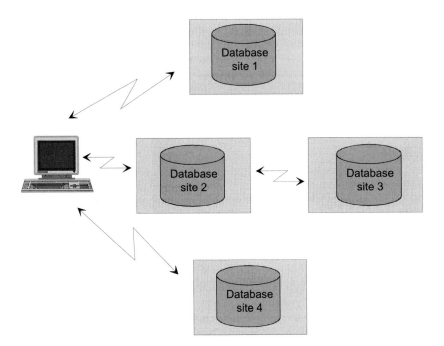

Figure 10.2 A single workstation or PC can access several remote database systems which may be located at different sites.

10.1.3 Interoperability

By interoperability, we mean that any computer could use data located on any other computer site by using software products located on another computer, as depicted in Figure 10.2. At the moment, distributed databases represent the first possible step forward fully interoperability.

10.1.4 Co-operative information systems

By co-operative information systems, we mean systems in which several people and machines work together.

10.2 Client-server architecture

Thanks to this architecture (Figure 10.3), it is possible to export some treatments and some data to different locations, instead of dealing with passive terminals able to execute several tasks or to store some data. And they can be linked to a data-server or an application-server. Both parts, client and server are independent but communicating: they are designed as separated

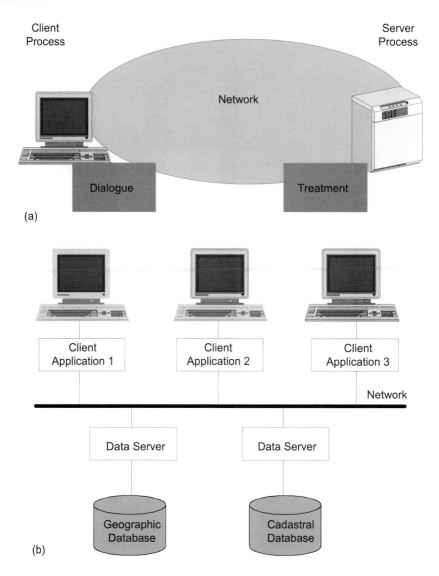

Figure 10.3 Principles of the client-server architecture. (a) The client process and the server process. (b) A small example in a planning context.

entities working in different contexts (computer, operating systems). The first part 'client' is generally installed on workstations whereas the server part is installed on the computer in charge of a special service. According to the service, the server can be a micro-, a mini-computer or even a mainframe. The client module is executed by the processor asking the service, that is to say taking charge, user dialogues, some checking and the routeing of the queries to the server. The 'server' module is installed on a different machine: it receives the queries, puts them into a queue, executes them in order and sends the results to the client. In other words, the server is in charge of several resources and typically of a database.

Generally speaking the language SQL (ANSI 1986) is used as a standard for communicating between the clients and the servers: the clients are sending SQL queries whereas the servers are sending relational tables back. At the moment, SQL does not incorporate spatial queries, the use of this language as a standard for distributed geographic databases is difficult. However when SQL 3^2 integrates spatial queries, it will be considered also as a standard in geographic multidatabase systems.

Advantages

The first advantage of task-sharing is obvious: each module can be designed separately in order to optimise its functions. The client-module can enhance the man-machine dialogue and the server can optimise access to the bases. A second advantage is the reduction of information volume which is exchanged between machines. The client sends queries to the server and the server sends results to the clients, possibly ornamented with acknowledgements. A third advantage is the flexibility. Indeed, the server module can be modified (for instance a new version of the operating system or a new version of the GIS) without changing the client module. Similarly, if the workstations are changed, for instance with better graphic possibilities, only the concerned modules must be changed.

Autonomy between systems is obtained thanks to standardised programming interfaces. The SQL language is typical for that purpose and it can be used as a protocol for exchanging data between clients and servers. For instance a client sends a SQL query and the server receives it, executes it and sends the resulting tables to the client. If SQL was the key-element of the first generation of client-server architecture, now the second generation is targeted to object-oriented systems based on CORBA.[3]

An example of the client-server architecture within a local authority is given in Figure 10.4. In the city-hall, a hub can link several departments namely Planning, Streets, Sewerage and Building Permits. This hub is linked to another hub allowing access to some other institutions linked with a municipality, e.g., the local water, electricity, gas and telephone companies, the Regional Administration, the Cadastre and the Geological Survey, etc. All computers are working either as clients or as servers.

---

2   See http://www.jcc.com/SQLPages/jccs_sql.htm
3   See for instance *Advanced CORBA Programming with* C + + by Michi Henning and Steve Vinoski (1999) Addison-Wesley Pub. Co, p. 1120.

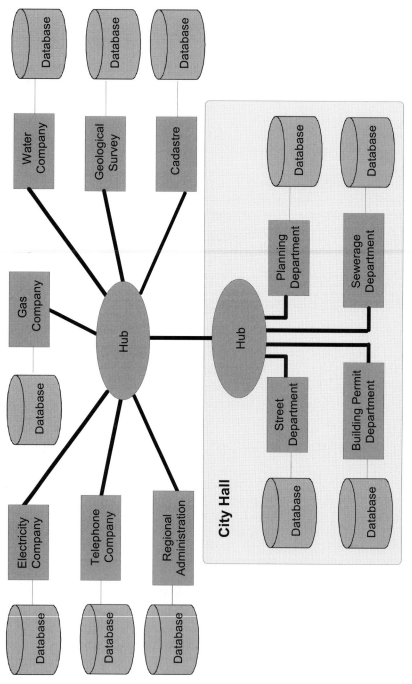

Figure 10.4 Example of a client-server architecture within a local authority.

10.3   Federating several database systems

In this section are successively presented generalities about multidatabases and finally some elements for the design.

10.3.1   Concepts and definitions

Here are definitions of some concepts which are intensively used in this domain. For more details, please refer to Oszu and Valduriez (1989), Bell and Grimson (1992) or Bobak (1993). Let us examine some of them.

Site

Usually a site possesses a single computer and a single database. Sometimes it can have several databases located on the same computer. Rarely, several computers are connected to appear as a single site from outside. Sometimes, a site is called a node in the network.

Distributed and federated databases

Concerning distributed and federated databases, several definitions exist. Generally speaking, distributed databases can mean two things. When the bases are carefully designed together, one speaks about distributed databases (narrow meaning), and when already existing databases must co-operate or interoperate together, one speaks about federated databases. But the expression distributed databases (wide meaning) can also represent distributed (at narrow sense) and federated.

Some databases can be tightly coupled, especially when the same DBMS or GIS is used. They are called loosely coupled especially when the links are made only for some dedicated applications. Several authors prefer the expression multidatabase, so referring to any kind of structure linking several databases together.

Homogeneous and heterogeneous distributed databases

Depending on the type of DBMS or GIS used, one speaks about homogeneous or heterogeneous databases. Homogeneous, when all databases systems are the same and heterogeneous when all database systems are different. Some authors are also more strict defining homogeneous databases not only running on the same DBMS but also having the same structure.

Data dictionary

By data dictionary, one means that a list is made including all data with their definition and structure. On each computer, there exists a local dictionary. At the global level, there exists also a global dictionary defining all data located on all sites. Seldom is data dictionary another name for metadata files (Section 6.8).

Local schema, global schema, import schema, export schema

In isolated databases, usually one defines a schema which represents the structure of a database; this schema will become a local schema in distributed databases. Due to some privacy reasons, a sub-part of this local schema is offered to the other members of the distribution/federation, and called **export schema**. In the reverse sense, one calls **import schema** the union of all export schemata coming from the other sites.

Now a single database user can indifferently work with the local schema and with the import schema. And in tightly coupled distributed database systems, the user must not distinguish when he is working locally or globally. By the expression 'external schema', one means the union of the local schema and the import schema as shown in Figure 10.5. By 'user schema', one means a subset of the external schema for a single user or application. The global schema refers to a schema storing the whole structure of all databases belonging to the distribution/federation.

Schema integration (Sheth and Larson 1990)

By schema integration, one means the process starting from the local schemata in order to synthesise them to build the global schema. The main difficulties to

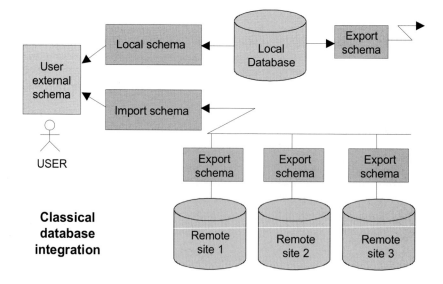

Figure 10.5 Relationships between schemata in multidatabases.

solve in this process are that the databases can have very different structures, the same data name can refer to very different things and so on.

Local and distributed queries

As said before, the user can work indifferently with the local schema and the import schema. One calls a local query when only the issuing site is activated to answer this query. By distributed query, one means that different sites (perhaps all) are activated in order to solve the query.

Horizontal, vertical and mixed fragmentation of tables

In relational database systems, it is usual to distinguish several kinds of fragmentation. One speaks about horizontal fragmentation of a relation when some tuples are located on a site and the other tuples on other sites. By vertical fragmentation, the table is split vertically, meaning that some attributes are located on a site and the other attributes on other sites. In this case, generally the keys are present on both sites so that the initial table can be reconstituted by a join (precisely an outer join[4]). For instance, a relation CUSTOMER can be split horizontally according to marketing sectors. Perhaps in another company, accounting information can be located in a site (say accounting department) whereas delivery information is another site (vertical fragmentation). In very special cases, mixed fragmentation refers to a mixture between horizontal and vertical fragmentation of the same initial relation. See Figure 10.6 for examples.

10.3.2   Data dictionaries

As defined before, the data dictionary integrates data semantics and location at global and at local levels. For each site, a local dictionary exists in which all database information is described (relations, attributes, views, etc.) by using widely metadata (see Section 6.7). At global level, similar information is regrouped into a global dictionary and additional information concerning location and fragmentation is also included.

A very difficult problem is the location of the global dictionary. A first idea is to define a privileged site to install this dictionary but in the case of a crash of this site, the system cannot continue to function. In order to palliate, the unique solution is to give each site a copy of the global dictionary. The drawback of this system is the necessity of maintaining their contents, especially during replicated global dictionary's updates. The consequence is

---

4   An outer join is a relational operation between two tables in which missing attributes are considered as unknown (null).

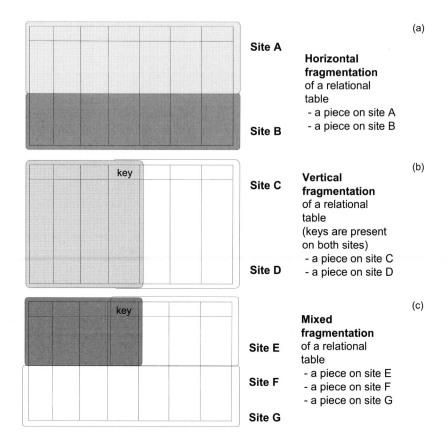

Figure 10.6 Horizontal, vertical and mixed fragmentation in relational databases.

the necessity of locking all sites during the updating of the global data dictionary.

10.3.3 Designing considerations

Two main cases must be distinguished during multidatabase designs, either one has to build a distributed system from scratch (top-down), or one has to federate different existing databases (bottom-up). Let us examine those two cases separately. Figure 10.7 depicts the differences between those cases.

Top-down approach

In this approach, one is starting from a global schema encompassing all data aspects of the systems. Then this global schema is split into different local

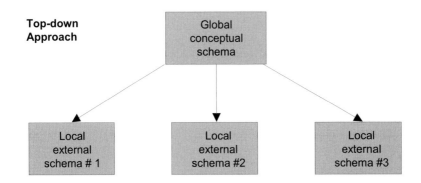

(a)　　**Decomposition of schemas according to several local dates**

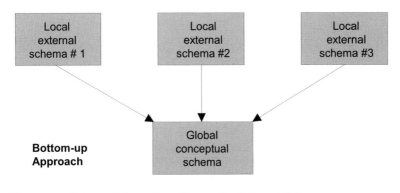

(b)　　　**Schema integration of several existing databases**

Figure 10.7 The top-down and the bottom-up schema designs. (a) Top-down or global-to-local approach for decomposing databases. (b) Bottom-up or local-to-global approach for integration.

schemata according to a placement strategy. By placement strategy, we mean that rules have been designed and followed in order to place adequately data fragments. For instance a city with several local districts can decide to put parcel records into the database of the concerned district, and to put in another computer site information regarding cultural affairs. In this approach, generally speaking, the same network, DBMS, operating systems and hardware are used, so giving what is called an homogeneous distributed system.

Bottom-up approach

The bottom-up or local-to-global approach is much more complex. The problem now is starting from different existing databases, to create a federation taking the maximum in common. Due to the variety of hardware, operating systems, data representation capabilities, it is often very difficult to federate different databases. Let us rapidly mention that in urban planning this is the more frequent situation. Indeed, all institutions or companies in charge of some municipal services have generally their own systems which are different and consequently a bottom-up approach of federating will be necessary

10.3.4 Spatial and thematic fragmentations

As said before, in relational databases, it is common to distinguish vertical and horizontal fragmentation or partitioning. But for geographic databases, this distinction apparently is not relevant; we prefer to speak about zonal fragmentation (also called spatial partitioning) and layer fragmentation (also called thematic partitioning) see Figure 10.8. Let us examine those cases.

Geographic partitioning or zonal fragmentation

By zonal fragmentation, we mean that the same information is split into different tables according to the geographic coverage, for instance each county can have its own information. In the top-down approach, this partitioning is generally very easy to perform. But in the bottom-up approach, due to inexact matching at the zone boundaries, some difficulties occur in order to ensure semantic, geometric and topological continuities between the different databases.

Thematic partitioning or layer fragmentation

By layer fragmentation, we mean that two different institutions can have different information on the same zone. For example, a cadastre layer, a building layer, a gas layer and so on. In the bottom-up approach the main difficulty is that some discrepancies exist between the positioning of reference objects (perhaps streets), implying problems similar to sliver polygons.

Heterogeneous fragmentation

In this case, several institutions having different kinds of information on different zones decide to federate their databases. It is not necessary to explain that this case is perhaps more common in geographic database federations and also the more difficult to deal with.

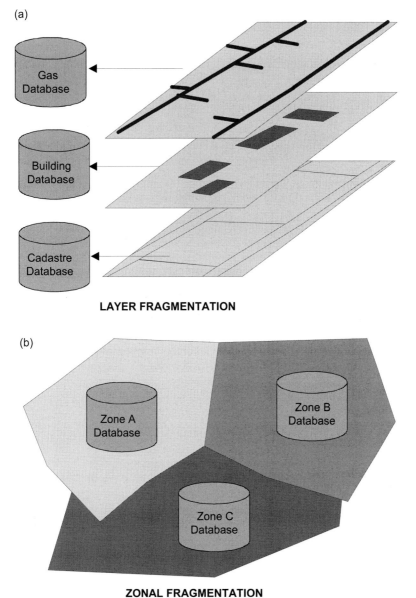

(a)

Gas
Database

Building
Database

Cadastre
Database

LAYER FRAGMENTATION

(b)

Zone A
Database

Zone B
Database

Zone C
Database

ZONAL FRAGMENTATION

Figure 10.8 Zonal and layer fragmentation in distributed geographic databases.
(a) Zonal or geographic partitioning. (b) Layer or thematic partitioning.

Fragmentation is interesting when designing a new geographic database and sharing information. In this case, we can suppose very easily that the same GIS is used by all partners (homogeneous GIS). However, when integrating existing geographic databases, the problems are totally different and solutions must be designed in order to solve those problems, coming essentially from inaccuracies in co-ordinates and more generally from the low quality level of some databases. Increasing quality or re-engineering some databases in connection with standards will be the key issues in order to get fully functioning geographic databases, distributions and federations.

10.3.5 Spatial schema integration

Once the network, hardware, operating systems and GIS differences are solved, the main problem is how to create the global schema taking all discrepancies into account. In this section, these discrepancies will be intensively explained with some hints in order to tackle them. But first, let us give some generalities about schema integration.

Generalities about schema integration

As previously said, generally speaking, export schemata represent a subset of the local schema essentially for privacy reasons. A possibility for schema integration is based on the ladder technique. In essence, it is a gradual approach integrating one schema after another in order to reach the final so-called global schema.

Difficulties to overcome

One of the difficult problems we have to face is that the remote data and the local data have several discrepancies and the mappings for export and import schemata must offer the necessary transformations at geometric and at semantic levels. During this effort, different kinds of conflicts must be solved essentially by using integration rules, namely semantic and structural conflicts. A first level allows the declaration of correspondences, then a second level allows the resolution of conflicts and then there is the real schema fusion. In addition to those conflicts, additional discrepancies exist for geographic information. Let us give some of them (Ram 1991, Kim-Seo 1991, Laurini 1994b, Laurini and Milleret-Raffort 1994):

- diversity of *spatial representations* of geographic information
- diversity of *global projections*
- diversity of *values* for the same item located in different sites

- diversity of *spatio-temporal sampling*
- variability of *definitions over time and space*
- discrepancies in *co-ordinate values*
- discrepancies in *boundary alignment*
- etc.

Solving those discrepancies will be a great challenge in the near future in order to design federated geographic databases. Let us analyse some geometric and semantic discrepancies with an example.

The Street Repair Department wants access to other homogeneous databases (that is to say with the same GIS or DBMS, for instance Oracle), the first one belonging to the water company and the last one to the gas company. Due to surveying errors, some geometric discrepancies will occur as shown in Figure 10.9.

The next problem is the diversity of spatial representations. Continuing our previous example, Figure 10.10 shows the structure of the different

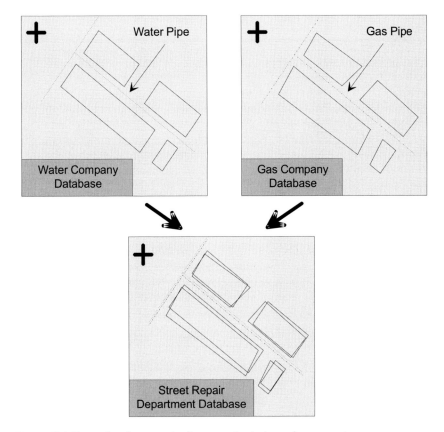

Figure 10.9 Example of geometric discrepancies in layer fragmentation.

Gas Company Database (G-site)	
G-STREET	(#street, street_name, (#axis_segment, width)*)
G-SEGMENT	(#segment, #point1, #point2)
G-POINT	(#point, x, y)
G-PIPE	(#edge, #node1, #node2)
G-NODE	(#node, x, y, z, type)

Water Company Database (W-site)	
W-STREET	(#street, (#right_segment, order)*, (#left_segment, order)*)
W-SEGMENT	(#segment, #from_point, #to_point)
W-POINT	(#point, x, y)
W-PIPE (#edge, #from_node, #to_node)	
W-NODE	(#node, x, y, z, (#edge)*, category)

Street Repair Department Database (SR-site)	
SR-STREET	(#street, street_name, (#parcel_segment)*,(kerb_segment)*)
SR-SEGMENT	(#segment, #point1, #point2, begin_address, end_address)
SR--POINT	(#point, x, y)
SR-G-PIPE	(#edge, #node1, #node2)
SR-G-NODE	(#node, x, y, depth, type)
SR-W-PIPE	(#edge, #node1, #node2)
SR-W-NODE	(#node, x, y, depth, type)

Figure 10.10 Example of semantic discrepancies in distributed urban databases.

databases stressing the semantic diversity of spatial representa-tions.[5] For instance, for the gas company, a street is defined by a unique set of axis-segments together with the street width whereas for the water company, we have two separated sets, one for the left side and another for the right side. Some of them require parcel segments and others kerb segments. The gas and the water companies have point elevation, whereas the Street Repair Department databases require point depth. Moreover the water and gas nodes can have different organisation of types: for instance, a tap is numbered 2 for the gas company and numbered 5 for the water company, whereas it is 1 for the Street Repair Department. It is also possible that the units of measurement are different. For instance, the gas company has metres and the water company decametres or feet.

---

5 Relations are written with the extended relational notation. See Section 2.10 and Laurini and Thompson (1992).

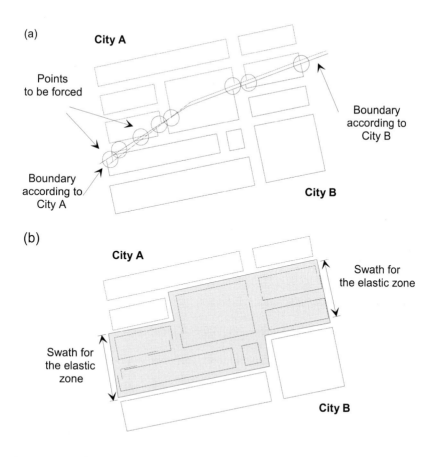

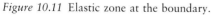

Figure 10.11 Elastic zone at the boundary.

Any co-ordinate in a geographic database has measurement errors. Especially at the boundary, two maps do not match. And this is the same for zonal fragmentation, the two database boundaries do not match. In order to correct this, some rubber-sheeting transformations must be launched with constraints in order to keep roads aligned, buildings rectangular and so on (Laurini 1996, Laurini and Milleret-Raffort 1995).

The solution of this problem is based on the consideration of not modifying the contents of other databases, but to solve this problem only when querying. A solution could be that when a database is newly integrated into the distributed system, a swath is defined along the boundary as illustrated in Figure 10.11 and the parameters of the rubber-sheeting transformation must be estimated and transmitted to the neighbours. For that, pairs of homologous points are defined. They are corresponding points belonging to each database and must be force-fitted in order to match the boundaries.

281

For defining this zone, a common width can be defined and used along the boundaries to match, for instance, 100 metres. Another possibility is to ask the user to give himself the swath limit, especially by means of a visual interface. For urban applications, it appears very interesting to use street axes as swath boundaries. Indeed, when delimiting arbitrarily, the swath limit can cut a building, a part of which must be transformed and another must not be transformed. By taking street axes, we eliminate this drawback (Figure 10.11). The only consequence is that some street limits are a little bit modified so that their parallelism is lost. In reality, this deformation is negligible. Whenever this deformation is too important, the user can take another street as the swath limit.

Boundary alignment

Figure 10.12 introduces very rapidly some problems to solve in the content matching of two databases: the boundary does not match, a building, a river and several roads are cut into two pieces, and a piece of a tributary is missing. Some solutions of the difficult problems can be seen in Laurini (1998). In urban information systems, this problem is common when integrating the database of an inner city with its suburbs.

Integration procedures

To conclude this aspect, when a new database is candidate to enter a federation, the procedure will be performed as follows:

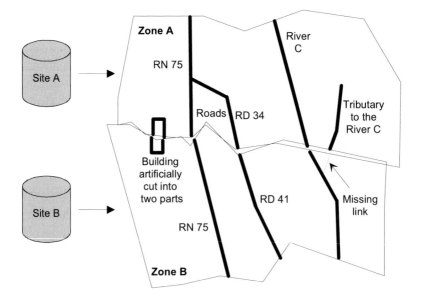

Figure 10.12 Examples of problems to solve in spatial multidatabase systems.

- implementation of mapping formulae from the database candidate co-ordinate system to the other, if necessary
- writing of export and import schemata
- delimiting outer swath
- providing homologous pairs via a visual interface
- selecting and preparing the elastic transformation for rubber sheeting.

10.3.6 Administration of spatial multidatabase for urban planning

The administration of a geographic multidatabase system is something important, since some consensus must apply between all the sharing institutions. Let us examine two aspects, the main objectives of the administration and the necessity of the creation of an inter-organisation protocol.

Administration of geographic multidatabase systems

Generally, each local information system has its own administrator. Let us call him, 'local administrator'. In addition, a 'global administrator' must be appointed in order to perform all integration procedures in liaison with the data custodians. Before introducing the tasks of the global administrators, let us resume the tasks that were supposed to be done by local administrators:

- design and implementation of the local conceptual database schemata, (logical database design)
- design and implementation of the local internal database schemata (physical database design)
- design and implementation of the external database schemata (local users views) in consultation with the local users
- implementation of local security and integrity constraints (when these are not explicitly included in the local schemata)
- definition and perhaps reorganisation of spatial index
- local database and performance monitoring
- planning and implementation of local database reorganisation as required
- documentation of the local database and local schemata.

Concerning the global administrators, their main tasks are as follows:

- definition of those portions of the local database which are to be contributed to the global database (i.e. local fragments of global relations)
- mapping rules between those fragments and the local data
- conversion rules between local and global data (e.g. miles to kilometres)
- specification of security and integrity constraints on the fragments
- documentation of the global data resource
- establishment and monitoring of standards for data naming, ownership, and so on

- liaison with global end users, management, data-processing personnel, network administrators and local administrators
- conflict resolution between different user/management groups
- definition of global security and integrity constraints
- development of a data resource plan
- definition of the global schema
- definition of the global dictionary
- definition of the global index
- definition of matching swaths and elastic transformation
- promotion of data sharing to avoid unnecessary duplication of data between local nodes and also to improve overall integrity.

Inter-organisational protocols

In addition to the appointment of a global administrator, the institutions sharing data within a multidatabase system must face several problems such as:

- copyright, indeed, each institution is owner of the data so every user must take this aspect into account when using data belonging to some other institution; one of the major difficulties in relying on updating. Suppose you realise that there is an error within the database of somebody else and you correct it. Who is the owner?
- access rights; not all end users are granted access or the use any kind of data everywhere, so some limiting access rights must be defined;
- difficulties during prototype implementation, indeed, during the integration procedure some sites can crash. Who is responsible?
- results property; suppose you create a map mixing information issued from different databases or sources. Who is the owner?

In order to solve those problems, the only solution is to set up or negotiate an inter-organisational protocol which has to be signed by all multidatabase partners. A nice possibility is to create a sort of agency in charge of enforcing this protocol.

10.3.7   Conclusions about federation

In urban planning contexts, a federation is very important because the planning process implies several institutions, each of them having their own databases and some connections with the other in order not only to share information, but also to share applications needing information in several databases.

In general, a database belongs to only one federation (Figure 10.13a) but in some cases a database can belong to several federations (Figure 10.13b). An example is given in Figure 10.14 in which a city database belongs to three federations:

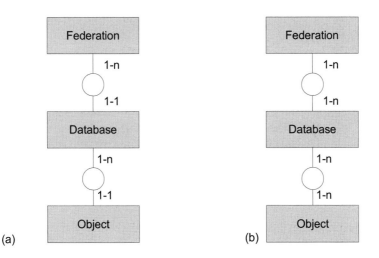

(a) (b)

Figure 10.13 Structure of federations. (a) Totally hierarchical, which is the more common case. (b) A database can belong to several federations, and an object can belong to several pages

- at local level, the federation of all administrations and companies in charge of local services such as water, gas, electricity, telecommunications, etc.
- at regional level, the federation of local authorities in order to foster information at an upper level of administration

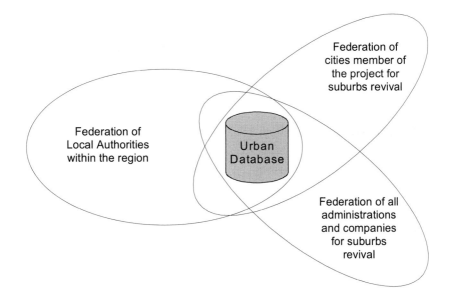

Figure 10.14 A city database can belong to several federations

- at nation-wide level, the federation of all cities which are members of a common national project, for instance the creation of a federation of databases related to the revival of their difficult suburbs.

10.4   Interoperability in geoprocessing

In Section 10.1.3, a wide definition and scopes of interoperability were given. Here we will address the special case of GIS interoperability and present rapidly the various ways to solve the problem (Vckovski 1998, Goodchild *et al.* 1999).

An open system (as opposed to a proprietary system) is one that adheres to a publicly known and sometimes standard set of interfaces so that anyone using it can also use any other system that adheres to the standard. As a consequence, one can plug in and play without difficulties. By interoperabilty, Webster's dictionary defines the word as the ability of a system to use parts or equipment of another system. The first aspect we have to mention is the existence of the OpenGIS Consortium (Buelher and McKee 1996) which regroups all major GIS vendors and was created for solving this problem.

As depicted in Figure 10.15, a sort of continuum can be found (Kucera and O'Brien 1997, Evangelatos 1999) to solve the problem of linking several computer systems for any organisation, from a stove-pipe system – that is to say, an *ad hoc* system – or to a really interoperable system, depending on the viewpoint, the quick solution for a single application or a sound system allowing the easy integration of a new application from the organisation point of view. We will finish this section by presenting a new approach for interoperability based on ontologies.

10.4.1   Necessity of a common exchange format

If we can consider n different GISs, we need to write $n \times (n-1)/2$ modules for adapting data; in addition, whenever a new GIS 'wants' to interoperate

	Short term solutions	Medium term solutions	Long term solutions
Concepts for solutions	List of virtual tables (schemata)	Manual mediators	Ontology based
Connectability	Stovepipe connectivity	Interconnectivity	Interoperability
Organizational aspects	Application view	Departmental view	Organization view

Figure 10.15 The implementation continuum, from the application view to the global organisation view.

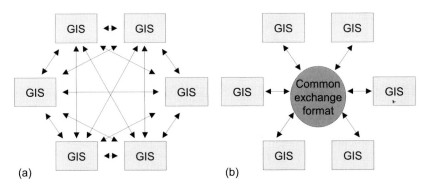

Figure 10.16 GIS interoperability. (a) Without a common exchange standard. (b) With a common exchange standard.

with the *n* previous ones, *n* new modules must be written. If a common exchange format is defined, only *n* modules must be designed and therefore when a new GIS format is considered, only one adapting module must be written. See Figure 10.16 for illustrating this aspect. So, the key idea of the OpenGIS[6] Consortium was to define this common exchange format, i.e. the Open Geodata Interoperability Specification (OGIS).

10.4.2 Introduction to OGIS for geodata interoperability

The OGIS is an attempt to provide a specification for a computing environment providing geodata interoperability by providing a framework for software developers to create software products that enable their users to access and process geographic data from a variety of sources across a generic computing interface within an open information foundation (Buelher and McKee 1996). It is based on two 'essential models' which describe the real-world situations and which are on the highest abstraction level, a feature and a coverage.

A feature is a representation of a real-world entity or an abstraction of the real world. Features have a spatio-temporal location as attributes and are managed in feature collections. The location is given by geometry and the corresponding spatio-temporal frame as a reference. Figure 10.17 shows the essential model of a feature with the object-oriented formalism as given in Chapter 2 with the following addition that a slash (/) on a line denotes a derived association.

A coverage is an association of points within a spatial/temporal domain to a value of a defined data type. The data type needs to be a simple type (e.g., numeric type) but can be any compound, complex data type. See Figure 10.18 for the essential model of coverages.

6   http://www.opengis.org

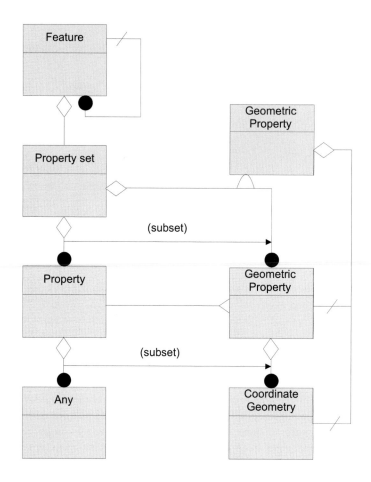

Figure 10.17 Essential model of OGIS feature type (Buelher and McKee 1996).

The scope of this section is not to detail all aspects on GIS interoperability. Interested readers can directly refer to the OpenGIS consortium WWW pages (http://www.opengis.org), or Vckovski (1998), or Goodchild *et al.* (1999). See also the proceedings of the Interop conferences (Vckovski *et al.* 1999)

10.4.3   *Implementing interoperability*

One interesting solution for implementing operability is to use mediators. By mediators (Figure 10.19), one means some pieces of software distributed along the network allowing the semantic matching of data (Wiederhold 1998, 1999). For instance, a very simple mediator can transform miles into kilometres,

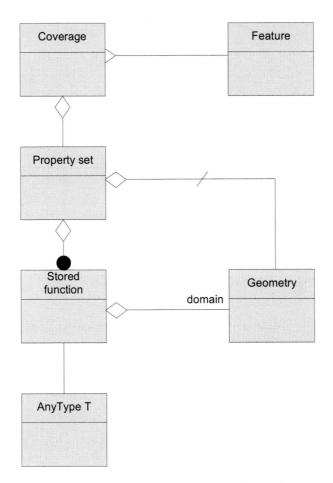

Figure 10.18 Essential model of OGIS coverage type (Buelher and McKee 1996).

another much more complex can transform month production into week production taking a whole calendar into account. By doing so, in the client-server architecture, a third layer is added, namely the mediators. Another interesting solution for implementing an efficient interoperable system is to base the architecture on existing standards. See Figure 10.20 for an example extracted from Evangelatos (1999).

Figure 10.19 Three-tier architecture for openness and interoperability.

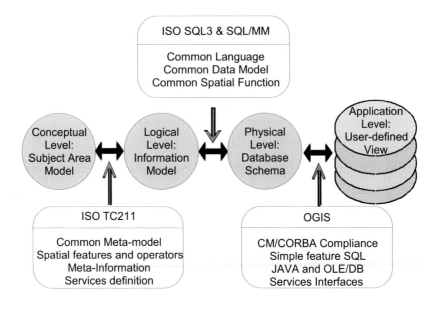

Figure 10.20 Linking standards in order to ensure interoperability (Kucera-O'Brien 1997, Evangelatos 1999).
Reprinted from Kucera, H.A. and Leighan, E. (1998) 'Mercator III: Toward a Canadian Geospatial Information Infrastructure', Technical Proposal Report to Natural Resources Canada, GeoAccess Division, March with kind permission of Henry Kucera.

10.4.4 *Ontology-based interoperability*

As presented earlier, the federation of different GIS requires first that any newcomer must be registered, and second that all schemata (import and export especially) have to be written. This solution is interesting when there is only a limited and fixed number of partners (say a few score). But when the number grows this rigid solution is not valid especially because it implies some reduction in the semantics. To alleviate this problem, a novel approach is based on ontologies. Remember that ontologies were defined in Chapter 6 as a formalised vocabulary to describe data. So the idea is to confer to all databases an ontology describing their contents, and to use those ontologies when exchanging or querying data. More explicitly, a shared ontology is defined as regrouping the vocabulary of all applications and co-operating databases. An architectural example is given in Figure 10.21 from Benslimane *et al.* (1999), in which we can see that an ontology is shared by different institutions.

A complete example of using ontology is given by Uitermark *et al.* (1999) for integrating data sets into the Dutch cadastre. Even so it is initially devoted to making several computers interoperable, it can give a very nice flavour of this technique (Figure 10.22).

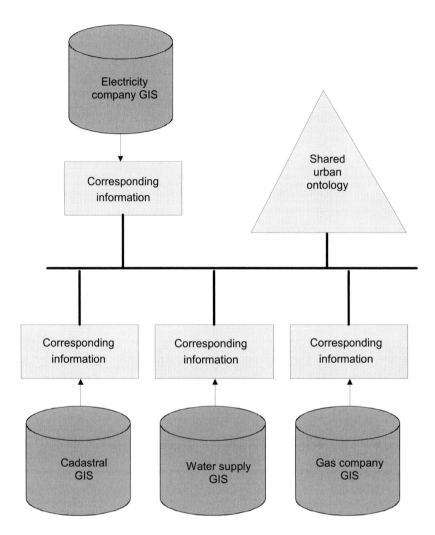

Figure 10.21 Architecture based on ontologies (Benslimane *et al.* 1999).

10.5 Architecture for groupware

As schematised in Chapter 8, the architecture of groupware or co-operative information systems is overall supported on software products which can be found very commonly in office automation for single- or multiple-connected users, such as tools for mailing, calendering, for project management and so on.

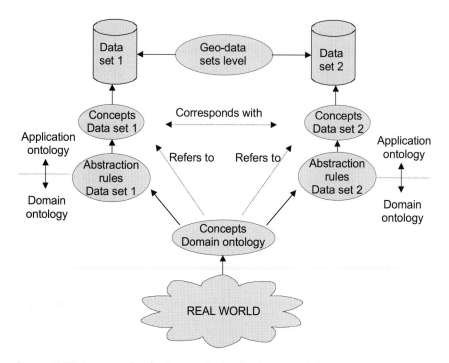

Figure 10.22 An example of using ontologies for interoperability. From Uitermark *et al.* (1999).

10.5.1 Generalities

Depending on the type of computer system used, CSCW tools, as illustrated in Figure 10.23 can be implemented in centralised approach, or in decentralised approach. In the decentralised approach, for a new user, the software architecture is more or less similar to the centralised approach, even if there are some major modifications due to the presence of the network and the communication. Indeed, one of the goals in a distributed approach is to be transparent to the users.

The architecture will be the following. Each member of staff will have their own computer with a local database. All computers will be linked by means of a network, and a central computer will act as a server storing all information needed to make the system work. An example for urban planning and municipal management is given in Figure 10.24. Regarding the contents of the local and global databases, information can be found in Figure 10.25.

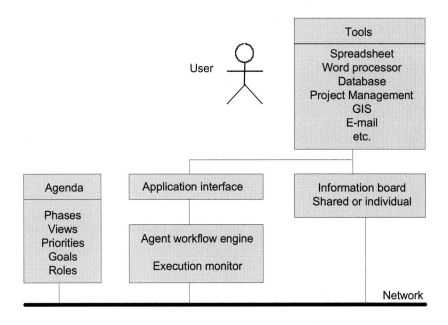

Figure 10.23 Architecture of a CSCW system.
After Mahling, Craven and Croft (1995).

10.5.2 Contents of the databases for urban planning

As previously said, there exists one global database and several local databases for running the co-operation. For instance, in urban planning the contents respectively of the global database and of the local databases are listed in Figure 10.25. However, the contents of the global database as given in this figure represents the staff of the urban planning department; for the other persons working in co-operation, the contents can differ according to their needs.

10.6 Datawarehousing

Among novel computer architectures, datawarehousing appears as a new possibility to organise and retrieve information. Since the inception of electronic data processing in the early 1960s or even before, organisations have created and maintained a lot of files. Initially on punch cards, then on magnetic tapes, and now on different kinds of disks, all these pieces of information were stored and sometimes put into archives when necessary. For decades, there has been practically no use for such data, except in very special conditions. Gradually the idea that this information was a sort of 'memory of the organisation' emerged, and the new possibility of computing allows analysis of these data, notably with

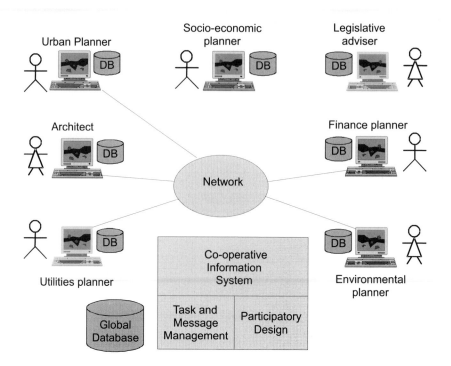

Figure 10.24 Architecture of a CSCW system for urban planning.

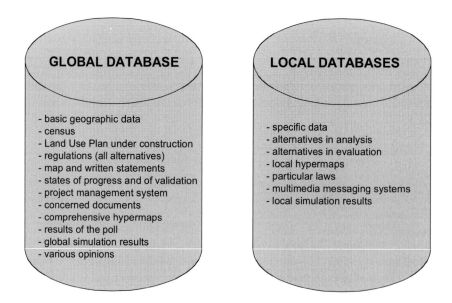

Figure 10.25 Contents of the central and the local databases for urban planning.

data mining (Westphal and Blaxton 1998) techniques in order to induce knowledge from existing files.

It is now, thanks to optical disks (CD-ROM) with a very high level of density, possible to store this information and to study it for finding data mining 'nuggets', i.e. new chunks of knowledge which can help decision makers and urban planners. In this section, the main principles of data-warehousing will be described. Then the data structure named as datacube will be detailed. Finally, some applications and potentialities in urban planning will be given.

10.6.1 Principles

The principle of datawarehousing is to consider all existing data within an organisation (not only current data but overall archives) and to make them accessible to potential users in order to help them in their work. In order to carry out this task, all existing data, generally with different formats are considered (Figure 10.26). By means of special tools for format transformation and integration, all data will be included in the datawarehouse. And after, the users can access these data for analysis or decision-making.

More precisely, some data were initially stored sequentially as flat files or with some other old data formats such as ISAM or VSAM, or with the relational model. Taking the variety of data formats into account, the tools for extracting and transforming these data are different. So, the datawarehouse consists of all these data after transformations. But the users are rarely interested in accessing data directly. What are of increasing interest are computer tools such as relational databases through client-server architectures, or perhaps tools known

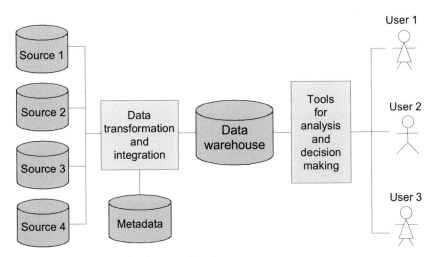

Figure 10.26 Structure of a datawarehousing system.
After Barquin and Edelstein (1997) with modifications.

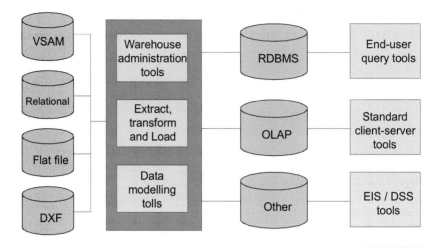

Figure 10.27 Components of a datawarehousing architecture system.
After Barquin and Edelstein (1997) with modifications.

as decision support systems (Turban and Aronson 1998), executive information systems (Watson *et al.* 1996) and so on. So the initial datawarehouse architecture must be transformed into a much more complex system as depicted in Figure 10.27. In this drawing, new software tools are mentioned, namely EIS and DSS. DSS (Decision-Support Systems) were rapidly described in the first chapter (Section 1.3), whereas EIS means Executive Information Systems, that is to say special software products targeted to executive officers. Among those special tools let us mention OLAP tools (Online Analysis Procedures) which appear very popular at the moment (Raden 1997).

According to Date (quoted by Barquin, p. 203), twelve properties are necessary in order to consider a new tool as an OLAP. Let us mention them very quickly:

1 multi-dimensional conceptual view
2 transparency
3 accessibility
4 consistent reporting performance
5 client-server architecture
6 generic dimensionality
7 dynamic sparse matrix handling
8 multi-user support
9 unrestricted cross-dimensional operations
10 intuitive data manipulation
11 flexible reporting
12 unlimited dimensions and aggregation levels.

Table 10.1 Comparing datawarehouses and operational databases

Datawarehouse	Operational database
Subject oriented	Application oriented
Integrated	Limited integration
Non-volatile	Continuously updated
Stabilised data value	Current data value only (generally)
Ad-hoc retrieval	Predictable retrieval

Having listed these properties, it is now possible to make a comparison between an operational database and a datawarehouse (see Table 10.1 for details).

One of the more important aspects of a datawarehouse is the way data is organised. For that a data dictionary is very important. More and more metainformation is used (see Section 6.7) to structure and to define data. With this new kind of computer architecture, according to (Bontempo and Zagelow 1998) datawarehouses can be considered as tools for transforming data, and especially old or ancient data into information, as explained in Figure 10.28.

10.6.2 Star structure and datacubes

According to Livingston and Rumsby (1997), the star schema articulates a design strategy that enforces clear and simple relationships between all the pieces of information stored in the datawarehouse. A simple star consists of a group of tables that describe the dimensions of the business arranged logically

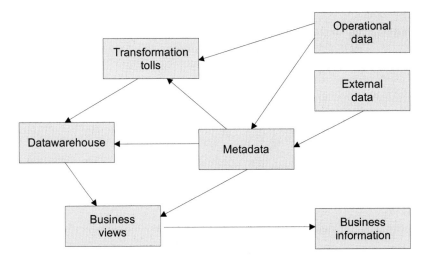

Figure 10.28 The process of transforming data into information by means of the datawarehouse technology.
According to Bontempo and Zagelow (1998) with modification.

297

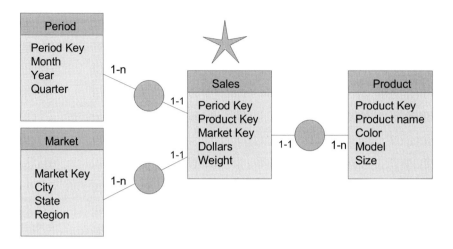

Figure 10.29 Simple star schema: one fact table (Sales), three dimension tables
(Product, Period and Market).
According to Livingston and Rumsby (1997).

around a huge central table that contains all the accumulated facts of the organisation. The smaller outer tables are the vertices of the star, whereas the larger table is the centre from which the vertices radiates. An example is marketing as given in Figure 10.29 in which the sales table is the centre of the star and the dimensions, the period, the market and the product.

The central table (the facts) contains millions of records. By themselves the facts might be meaningless, but analysts use them in conjunction with constraints on the dimensions to run standard calculations as counts, sums, averages and so on. If we follow this example, other common dimensions could be customer, store, vendor, demographics, promotions and so on. Another very popular way of organising warehouse data is to use the concepts of multidimensional datacubes. For instance, in a company, sales information can be organised along three dimensions; time, geography, items.

Figure 10.30 tries to explain the datacube concept taking an example from building permits. Figure 10.30a illustrates the three selected dimensions, namely the district number, the constructed floorspace and the time period. Each cell of this cube represents a number extracted from archives as given in Figure 10.30b. Starting from this structure, several operations can be made, and only two are depicted in Figures 10.30c and 10.30d, one time-independent, and the other zone-independent.

OLAP tools consist of a toolbox for analysing data based on datacubes in order to find some regularities. The conventional OLAP operators are split, merge, roll up, drill down, dice and slice.

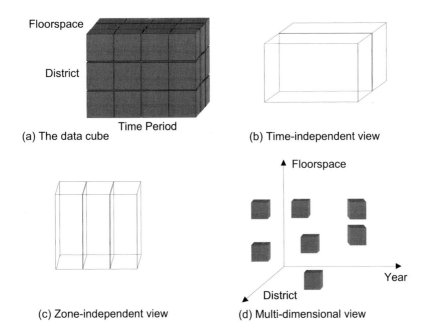

Floorspace

District

Time Period

(a) The data cube

(b) Time-independent view

(c) Zone-independent view

(d) Multi-dimensional view

Floorspace

Year

District

Figure 10.30 Example of a datacube for accessing a datawarehouse for floorspace analysis. (a) The datacube. (b) A time-independent view. (c) A zone-independent view. (d) The multi-dimensional view.

Datawarehouse Application Menu

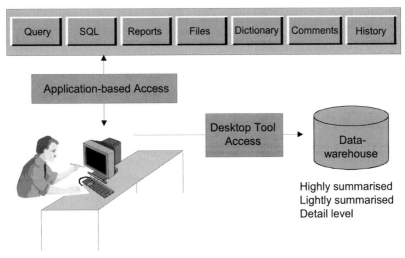

| Query | SQL | Reports | Files | Dictionary | Comments | History |

Application-based Access

Desktop Tool Access

Data-warehouse

Highly summarised
Lightly summarised
Detail level

Figure 10.31 Example of a datawarehouse implementation.
From Barquin and Edelstein (1997), with modification.

10.6.3   Interface

Figure 10.31 from Barquin and Edelstein (1997) represents an example of a datawarehousing interface, in which we can see the menu and the main components for accessing and summarising information from dataware-houses.

10.6.4   Datawarehouse architectures (Bontempo and Zagelow 1998)

Data in a datawarehouse should be reasonably current, but not necessarily up to the minute, although developments in the datawarehouse industry have made frequent and incremental data dumps more feasible. Data marts are smaller than datawarehouses and generally contain information from a single department of a business or organisation. The current trend in data-warehousing is to develop a datawarehouse with several smaller related data marts for specific kinds of queries and reports.

Datawarehouses can be configured according to various architectures:

- centralised datawarehouse, which is a very simple architecture
- datawarehouses and data marts; in this case a data mart typically contains a narrower scope of data characterised by a single subject so several data marts can be linked to a centralised datawarehouse
- distributed datawarehouse in which several datawarehouses are con-nected via a network; such distributed or virtual warehouses require strong capabilities for distributed management providing at the minimum a global view to potential users.

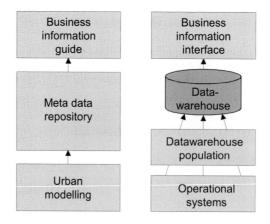

Figure 10.32 Architecture of a datawarehouse system.
According to Devlin (1997) with modifications.

Client Components

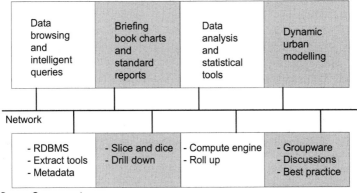

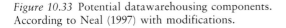

Server Components

Figure 10.33 Potential datawarehousing components.
According to Neal (1997) with modifications.

An example of using a datawarehouse for urban planning is given in Figure 10.32, in which the warehouse is populated by data coming from different operational systems, in parallel with a meta-data repository explaining the semantics of the stored data.

More complete is the following drawing (Figure 10.33) giving a list of tools which can be used for data mining in the datawarehouse in a client-server architecture.

10.6.5   Methodology

Gardner (1998) has recently proposed a complete methodology in order to design a datawarehouse (Figure 10.34) in three steps:

- datawarehouse planning
- datawarehouse design and implementation
- datawarehouse usage, support and enhancement.

Then these three steps are disaggregated into smaller phases, in which the last steps present a general feedback for enhancing the whole system.

10.6.6   Datawarehousing and urban planning

To conclude this section, let us present very rapidly an example in urban planning concerning building permits. The relationships can be as follows:

```
Building-Permits ( Bp-ID, Plot-ID, permit-date,
                   Owner-ID, building-type, floorspace)
Location (Plot-ID, district-ID, area)
```

301

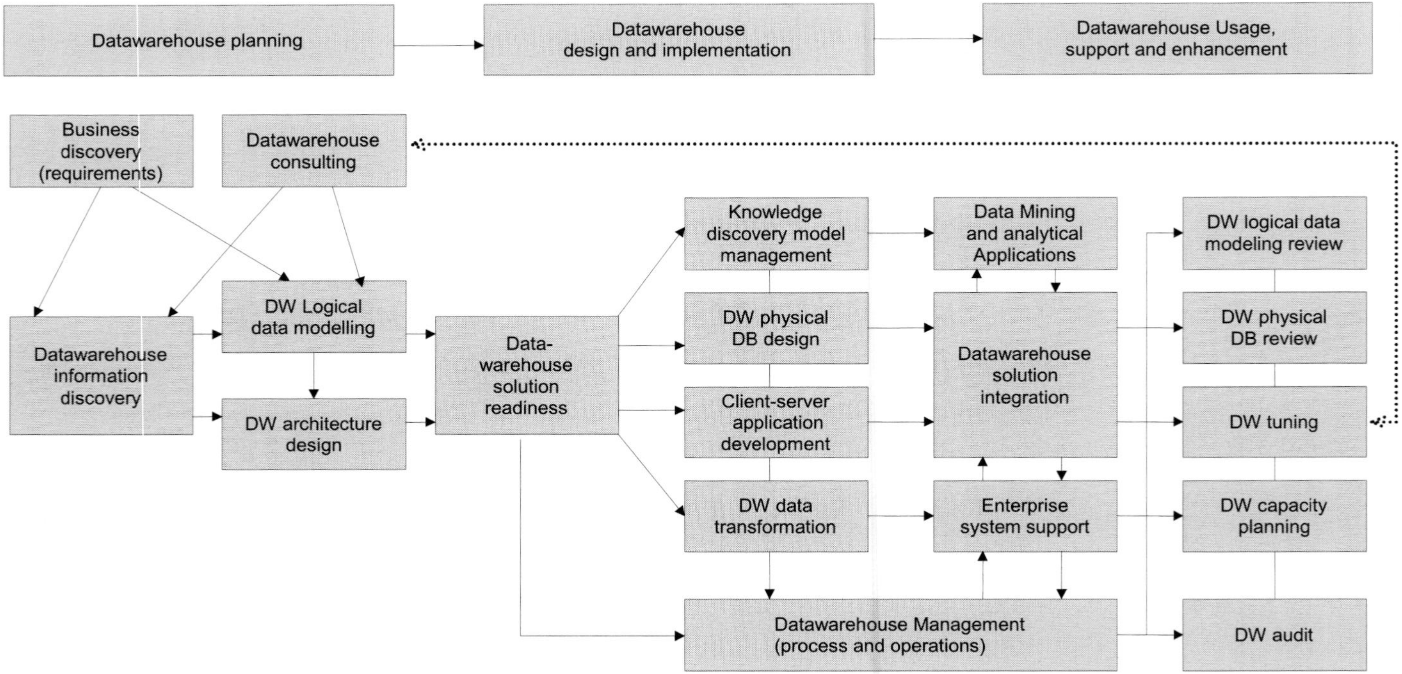

Figure 10.34 Building and managing a datawarehouse.
According to Gardner CACM Sept 1998 p. 57 with permission.

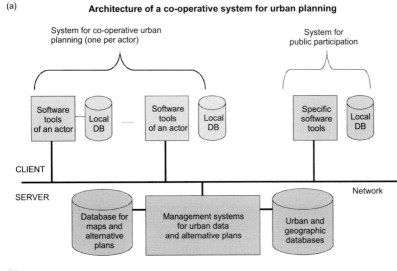

(a)

Architecture of a co-operative system for urban planning

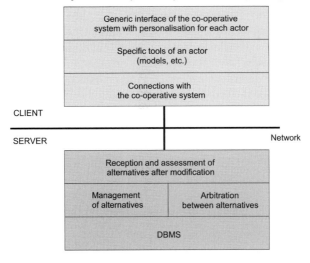

Figure 10.35 Architecture of a co-operative system for urban planning. (a) Principles. (b) Inner layers.

```
Transaction (Trans-ID, Plot-ID, date, seller-ID,
                buyer-ID, transaction-date, amount)
Completion (Bp-ID, completion-date, modification)
```

Starting from those relationships, it is easy to make cubes such as:

```
Cube (Cube-ID, floorspace, plot-ID, completion-date)
```

Thanks to this structure, data will be examined by data mining techniques in order to study several aspects such as:

- the evolution of land consumption
- the evolution of prices
- the spatial distribution of new constructions
- the chain effects between buildings, services,
- etc.

10.7   Conclusions

As related in the introduction to this chapter, the goal was to give the reader some elements in order to understand the evolution of computer techniques. Decades ago, the structure of information systems for urban planning was very simple: a mainframe, some disks, a few screens, several printers, several plotters, sometimes a few digitising tables. Now, with the advent of the Internet, the systems are much more complex, allowing several planners to work together in a groupware context, allowing citizens not only to be informed, but also to visualise it with virtual reality and also to make themselves simulations of the consequences of the plans (see Figures 10.35a and b for details). All those aspects imply also telecommunications, interoperability between several systems, sometimes in real time by using satellites, and will be described in the next chapter. Following the evolution of the last decades, I hope that novel tools will be created in order that any citizen can co-operate with the local authorities to create a more human city of the future.

11

REAL-TIME INFORMATION
SYSTEMS FOR URBAN
ENVIRONMENT AND RISKS
MONITORING

Environmental planning is becoming more and more important in urban planning. In the previous decades, urban environment was only studied in differed time; but now, thanks to telecommunications and real-time sensors, environmental monitoring can be carried out in real time. The objective of this chapter will be to examine the objectives and the architecture of those new computer systems.

Linked with the environment is the problem of crisis especially when the city can be subject to natural and technological risks. In other words, for centuries disaster mitigation was commonly included in urban planning practice, for instance, trying to avoid floods by constructing new dams, now the continuous monitoring of the environment variables allows more easy analysis, forecasting and prevention.

This chapter is split into two parts. The first examines the architectures of real-time decision-support systems for environmental modelling, for risk management and for humanitarian assistance in cities. The second part describes a generic structure for telegeomonitoring.

11.1 Real-time systems for environmental monitoring

For decades, conventional urban information systems were generally made for municipal office applications. But novel applications, especially for environmental management, imply the necessity to work in real time. Applications for air pollution or river control fall in this category. In addition, we can mention traffic control, and the prevention of flood and of any kinds of urban risks. In this section, we will introduce the reader to the architecture of such systems which are becoming increasingly common. Now, they can be regrouped under the umbrella of telegeomonitoring systems (Laurini 1999).

11.1.1 Example for air pollution control

Figure 11.1 depicts the architecture of a system for the monitoring of air pollution in a city. Various sensors are distributed all over the city. These sensors

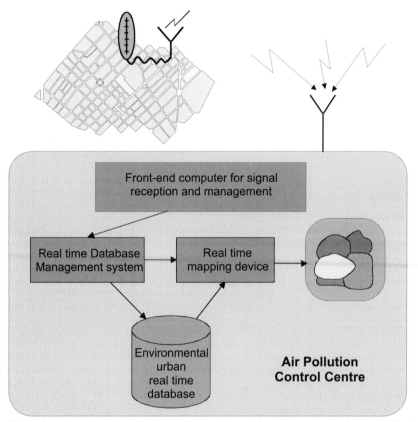

Figure 11.1 Functional architecture of a real-time information system for air pollution control.

measure some parameters and are linked to a special communication system in order to send data and receive orders. Among others, an interesting possibility is to use cellular phones as transmission devices. Moreover this figure also gives the sketch of the complete system which is linked to a real-time mapping device giving an animated cartography of air pollution. The sensors are, for instance, initially set to send information every hour. If necessary, the control centre can call them in order to modify the period of data sampling.

11.1.2   River flood and pollution control

For monitoring a river, and especially a river with several polluting companies, sensors are settled along the river and along its tributaries. Generally these sensors have to be located downstream of major hazardous plants, such as chemical or nuclear which are likely to pollute. These sensors can measure

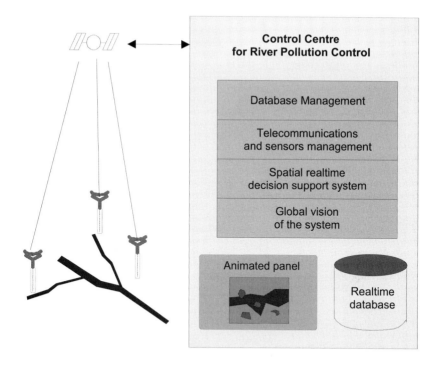

Figure 11.2 Sketch of a computer system for river pollution monitoring.

some chemical, physical and biological parameters and regularly send information to a control centre as sketched in Figure 11.2. Interesting systems have been devised for flood prevention; let me mention the State of Parana in Brazil (Cortopassi-Lobo and Guetter 1999). Some other systems can be found for seismic or volcano monitoring.

11.1.3 Mobile systems

In the previous examples, the sensors were fixed. But for some applications, the sensors move. Even for air and river pollution, we can imagine cars or boats equipped with sensors measuring dynamically. An example of a computer system based on measuring boats navigating along the lagoon of Venice can be seen in Cambruzzi *et al.* (1999).

Figure 11.3 describes, in a simple and informal way, the proposed operating and functional mechanism to support the decisions and the correcting measures in case of detection of anomalous or dangerous environment conditions:

1 an anomalous situation appears (e.g. anoxic conditions, dispersion of pollutants, biological proliferation)

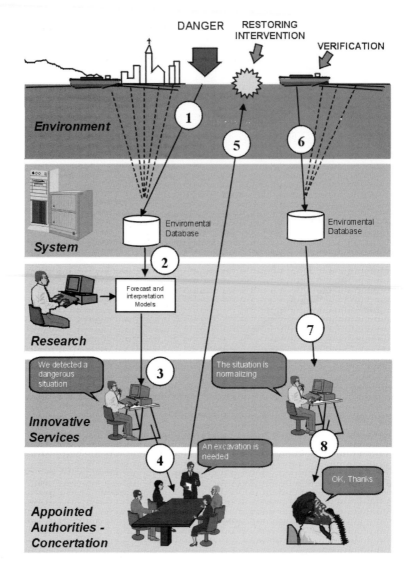

Figure 11.3 Description of the system for the monitoring of the Venetian lagoon (Cambruzzi *et al.* 1999).

2   the collected data are automatically compared with the allowed threshold values
3   the service operator detects the event, collects available data and transmits them to the appointed authorities
4,5 the authorities decide on a particular restoring intervention, that is timely performed

6   the acquisition of environmental data continues day by day; they are used to verify the obtained results

7,8 the results of the intervention are resumed and forwarded to the appointed control authorities.

Here the on-board computers equip those vehicles, together with some telecommunication systems. Some other examples can be found for the police, civil security, firemen, etc. Remember that a mobile vehicle system for data acquisition was described briefly in Section 3.7 based on sensors linked to voice data.

11.2   Towards telegeomonitoring[1]

As presented in the examples depicted in the previous paragraphs, for several years, new geographical and urban applications are emerging in which the characteristics of communications and real time have been very important. In other words, we do not have to deal with applications for which conventional cartography was the target, but in which spatial aspects are involved in management in real time or with very strong temporal constraints, such as mission-critical applications. For this kind of problem, the intensive usage of position systems (such as GPS) is the key element allowing the vehicles to know their positioning and to exchange any kind of information by any kind of telecommunication means. For instance, the management of a fleet such as for the police or for rapid delivery needs to know at all times the exact position of all vehicles. By means of GPS and embarked or on-board computers, they can communicate with some control centre, which in turn can send them other information. In this family of applications, the control centre is equipped by huge electronic panels, which represent the moving objects' locations and trajectory superimposed over some base map.

In order to fulfil this task, the vehicles must have on-board computers, connected to GPS and to the control centre. In some applications, the control centre does not exist, and all vehicles are exchanging only between themselves information regarding their positioning. So, we can see novel applications, engendered by geoprocessing and telecommunications, all possessing common characteristics and implying adapted researches. We think that we are facing a new discipline called telegeomonitoring (Laurini 2000). Telegeomonitoring, as a child of Geographical Information Systems and telecommunications, can be considered as a new discipline characterised by positioning systems, cartography, the exchange of information between different sites and real-time spatial decision-making.

---

1   Some elements of this section are from Laurini R. (2000) *An Introduction to TeleGeoMonitoring: Problems and Potentialities.* GIS Innovations, edited by P. Atkinson and D. Martin, Taylor and Francis 1999, pp. 11–26.

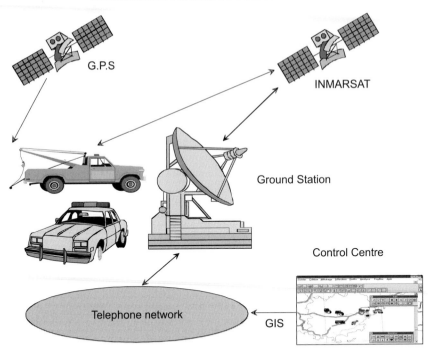

Figure 11.4 Schema of a system for the monitoring of toll motorway traffic (Tanzi *et al.* 1998, Laurini 1999).

Applications ranging from traffic monitoring, fleet management, to environmental planning, transportation of hazardous materials and surveillance of technological and natural risks, have all in common not only some functional similarities, but also several computer architectural aspects such as centralised, co-operative or federated. A description of those applications will be made in order to compare them. In addition, some generic problems arising in the interoperability of several telegeomonitoring systems will be addressed.

In Figure 11.4, the schema is given showing the architecture (Tanzi *et al.* 1997, 1998, 1999) in which we can see that any motorway vehicle is linked to two satellites (GPS and INMARSAT[2]) in order to communicate not only their position, but also to exchange any information with the control centre. So, telegeomonitoring can be seen as an extension of GIS by adding some telecommunication aspects. In reality, we think that it is more linked to real-

---

2   INMARSAT is a satellite system for telecommunication covering the whole earth. Application is primarily maritime communication, although it is hoping to expand more into personal communications with a newer technology, Inmarsat 3. It is comprised of four geo-synched communication satellites for two-way communication. See http://www. inmarsat.org/

time spatial decision support systems. Now, we can define telegeomonitoring as 'a new discipline trying to design and structure geographical information systems functioning in real time with sensors and on-board or mobile components, exchanging data in order to allow short-term and long-term decision-making based on data regularly received in real time by any means of telecommunications'.

11.3 Computer architectures

After the description of several cases, some common characteristics can be distinguished, such as using GPS, data exchange, existence of a real-time database, existence of fixed sensors, mobile or on-board computers, control centre, decision-support system, anticipation by simulation, decision, animated cartography, etc. We can see that these common aspects are at functional or architectural levels.

All these previous applications have in common a key component named control centre, and some mobile components with on-board computers using differential GPS. Sometimes some fixed sensors are also present. So, we can distinguish passive vehicles which can emit and receive orders, and intelligent vehicles which can modify their path according to some local or global criteria. Similarly, one can imagine also intelligent sensors which can change their strategy of acquiring data according to the environment.

Generally speaking, we can consider two types of sensors, fixed sensors and on-board sensors. Indeed, in addition to GPS receivers (which can be considered as special sensors), vehicles can possess cameras or other types of apparatus, for instance to measure external temperature. In the sequel, both sensors and on-board vehicles will be integrated into the concepts of vehicles, and only the notion of isolated sensors will be used in this chapter.

The control centre and the other components exchange data which are multimedia and coming from different sources by using extensively wireless telecommunications, perhaps with several satellites. All these pieces of information will be stored in a real-time geographical database from which some other more elaborate information can be derived.

Another way to classify telegeomonitoring systems is to take the number of persons in manned vehicles, namely, one-man vehicles, or multi-man vehicles. When there is only one pilot, the program is much more complex than when there is a co-pilot. Indeed, when the pilot is by himself, it is difficult for him both to drive the vehicle and to interact with the computer, whereas when there is a co-pilot, this person can be in charge of the computer. For one-man vehicles, an interesting solution is voice-based cartography replacing conventional graphic cartography by the emission of geographic information by means of a vocal device. The important consequence is that a one-man vehicle solution is much too complex at the moment from a computer point of view.

11.3.1 Models of architecture

Bearing all that in mind, it is possible to distinguish three architectures, centralised, co-operative and federated. We shall now examine them in detail.

Centralised architecture

The centralised architecture (Figure 11.5) shows the existence of a control centre where data converge, and which sends data and statements both to mobile vehicles and fixed sensors. Let us mention that in this architecture, only the control centre has a global vision of the system at all times. Of course, a different copy can be sent to the other components when necessary. The great weakness of this architecture is that in the case of a crash of the control centre, this telegeomonitoring system does not function any more.

When a new vehicle enters this system, it informs the control centre which in return sends adequate information. Among operations at the initialisation, the control centre must update the local database so that the new vehicle can function efficiently. Let us remark that at the moment, this architecture is the more common. In order to alleviate crashes, a solution is to create a mirroring control centre; this solution is often taken for crisis management teams.

Co-operative architecture

In the co-operative architecture (Figure 11.6), that is to say without any central site, all fixed and mobile components exchange information between themselves. As a consequence all sites have a global vision of the context.

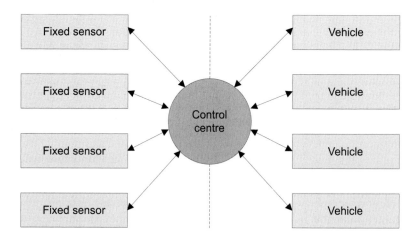

Figure 11.5 Centralised telegeomonitoring architecture.

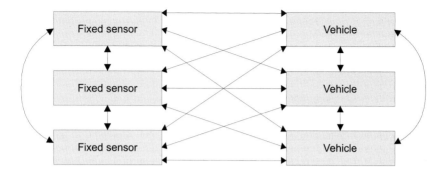

Figure 11.6 Co-operative architecture for telegeomonitoring.

When a new vehicle enters the system, it informs all vehicles which when necessary will send it information.

This architecture is much more robust to crashes than the centralised architecture. On the other hand, the quantity of information to be exchanged is much more important. Indeed, one of the priorities is the updating of all databases; if one copy is different from the others, some difficulties, sometimes drastic, can occur. In those cases, similar problems as those encountered in distributed database systems exist. However, the more serious case is called the Byzantine situation in which one or several components are either defective or always sending erroneous messages, especially due to the drifts of the sensors (see for instance Simon 1996 for more details). In this crucial case, the defective components must be detected as early as possible, and the messages must be corrected. Another problem is the authentication of received messages. Think for instance of the hijacking of a lorry transporting hazardous materials. In order not to attract attention from the control centre, the hijackers will send innocent messages. Of course, a control centre soundly designed must diagnose this case.

Federated architecture

Some vehicles can be connected to several telegeomonitoring control centres. For instance a lorry transporting petrol must be in connection with its haulage contractor, the motorway system, and a more general system for hazmat surveillance. In this case, the lorry must have three distinct on-board computers, or a single one implying problems of interoperability between the three systems?

As an example, Figure 11.7 depicts the case of a vehicle belonging to three telegeomonitoring systems (Figure 11.7a). From a general point a view, one telegeomonitoring system consists usually of several vehicles, and one vehicle can belong to several systems (Figure 11.7b). First, the vehicle (Figure 11.7c) belongs

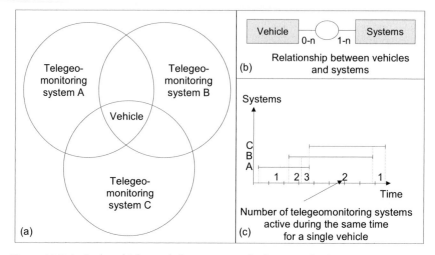

Figure 11.7 A single vehicle can belong to several telegeomonitoring systems. (a) Ownership diagram. (b) Relationship between vehicles and telegeomonitoring systems. (c) Temporal sequence.

to system A, then to system B, and finally to C; then it exits from A, then from B, and so on. Therefore, we can see that, according to the number of systems and their nature, the kinds of interoperability problems will differ. As a consequence, we will state that the belonging can be successive, for instance for aeroplanes which pass successively from one air traffic control tower to another one, or simultaneously, as in the example presented in the previous section.

11.3.2 Functional aspects

From a functional point of view, we can state that these applications can be seen as variants of real-time spatial decision support systems, in which information pieces come from sensors and on-board components. In order to assist short-time decision making, large-format screens can be used, based on animated cartography. Or more exactly, animation is driven from the received data. For the long term, numeric simulations must be performed starting from data stored in the control centre, and some forecasting models. So the consequences of a decision will be estimated.

An example of functional architecture of such a real-time decision-making support system and of its environment in given Figure 11.8 (Tanzi *et al.* 1998). On the other hand Figure 11.9 details functional aspects of both central and on-board sites.

Remember that there exists a substantial literature on interactive decision support systems, but it is much more limited with respect to real-time and spatial DSS (Section 1.3); in addition, to our knowledge, little research has been done on cognition for real-time systems.

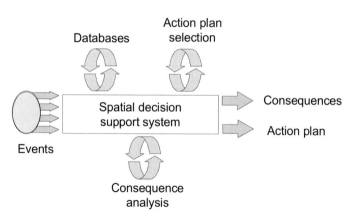

Figure 11.8 Environment of a real-time decision support system (Tanzi *et al.* 1998).

11.3.3 Comparison

Starting from these studies, a comparison can be done (Table 11.1) emphasising common issues and differences between several classes of applications, from a functional and an architectural point of view.

11.4 Urban major risk management

According to Cochran and Power (1999), the social and economic impacts of major disasters caused by storms, floods, landslides and earthquakes have created pressure to apply new and emerging technologies to improve the response and co-ordination of efforts in support of disaster[3] situations, especially in cities. Both domains, urban planning and disaster management have a lot in common, and the scope of this section is to make the reader aware of the specific problems that urban planning and monitoring must include to deal with disaster management, especially in the early phases. In other words, what is sometimes called 'emergency planning', namely the late phases of urban risk management is outside the scope of this chapter. For that purpose, there is a vast array of volunteer, industry and government organisations involved in the delivery of aid and assistance in disaster situations. This network of actors requires a large effort to plan and co-ordinate activities in advance of impending disasters, during the event and in many cases for extended periods of time after the event has passed (Collin 1995). In order to reach those objectives, some special computer, telecommunication and information systems must be designed.

---

3   For a natural disaster bibliography, please see http://www.colorado.edu/hazards/.

Table 11.1 Common characteristics of some classes of telegeomonitoring systems

Activities		Examples of applications				
		Motorway traffic management	Fleet management	Hazmat transportation	River pollution monitoring	Risk monitoring
	Use of GPS	Yes, on some vehicles	Yes, on all vehicles	Yes, on all vehicles	Possibly for some sensors	Possibly for some sensors
Computer architecture	Fixed sensors	Yes	Possibly	Possibly	Yes	Yes
	On-board components	Yes	Yes	Yes	Generally no	Generally no
	Real-time DB	Yes	Yes	Yes	Yes	Yes
	Data exchange	Yes, between vehicles, sensors and control centre	Yes, between vehicles, sensors and control centre	Yes, between vehicles, sensors and control centre	Yes, between all components	Yes, between all components
	Control centre	Yes	Yes	Yes	Yes	Yes
Functional architecture	Real-time decision support system	Yes	Yes	Yes	Yes	Yes
	Anticipation by simulation	Yes	Yes	Yes	Yes	Yes
	Animated cartography	Yes	Yes	Yes	Yes	Yes

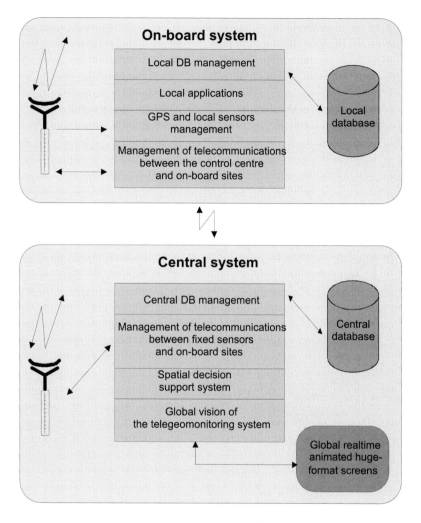

Figure 11.9 Functional details for central and on-board systems.

This section is organised as follows; first the main phases or the management of urban risks will be examined, then a wide specification list of information required for those tasks will be presented very rapidly in order to give a first attempt at the structuring of a possible system. Based on these elements, the connections between GIS and urban risk mapping will be given. We will conclude this section by offering firstly the structure of an extranet targeted to citizens and dealing with urban areas' natural risk management and control. The next section will detail an example of telegeoprocessing for technological risks in cities.

Table 11.2 Phases for disaster management systems in cities (FEMA 1998)

Mitigation	Mitigation is the process of taking sustained actions to reduce or eliminate long-term risk to people and property from hazards and their effects.
Preparedness	Provide the leadership, policy, financial and technical assistance, training, readiness, and exercise support to strengthen (a) community readiness through preparedness and (b) the professional infrastructure of trained and tested emergency workers, community leaders, and public citizens who can prepare for disasters, mitigate against disasters, respond to a community's needs after a disaster, and launch an effective recovery effort.
Response	Response is the process of conducting emergency operations to save lives and property by positioning emergency equipment and supplies, evacuating potential victims, providing food, water, shelter, and medical care to those in need, and restoring critical public services.
Recovery	Recovery is the process of rebuilding communities so individuals, businesses, and governments can function on their own, return to normal life, and protect against future hazards.

11.4.1   Main action phases

Disaster management systems are characterised by the information which is required to support decision making in all chronological phases of a disaster – mitigation, preparedness, response and recovery. These phases are defined by the Federal Emergency Management Agency,[4] (FEMA 1998) as follows (Table 11.2).

11.4.2   Information requirements

Prevention or mitigation of disasters is a strategic activity planned long before disasters may occur. Preparedness can be both strategic and tactical through pre-deployment and readiness of resource in anticipation of a disaster. Response is a tactical activity requiring rapid acquisition of information. Recovery contains both short- and long-term information requirements.

All disaster information can be characterised in terms of timeliness, consistency, comprehensibility, accuracy, and flexibility (Farley and Hecht 1999). The degree to which these characteristics must be addressed is dependent upon the characteristics of the disasters and natural or tech-nological risks for which decisions must be made. Each disaster can be characterised in terms of magnitude, urgency, infrequency and unpre-

---

4   See http://www.fema.gov/

dictability, multidimensionality, geography, social and political environment, and after-effects.

According to Cochran and Power (1999), the goals of disaster information have been identified by the National Research Council of Canada as follows:

- to improve decision making before, during and after emergencies through improved access to and quality of information
- to provide information products that are specifically designed to meet the needs of users
- to promote efficiency and cost effectiveness
- to stimulate and facilitate mitigation.

Table 11.3 (partly from Cochran and Power 1999) identifies the data requirements to build such a system. Once the data are selected, an interesting idea is to add the corresponding metadata. For the example of the San Francisco bay risk management system, visit the site of the metadata for risk at the following address: http://www.regis.berkeley.edu/glinks/mdata/bayrisk94.html.

Table 11.3 Information resources for decision making for urban risks. From Cochran and Power (1999) with modifications

Base data	Topography; political boundaries; public land survey system; geographic names; demography; land ownership/use; critical facilities; etc.
Scientific data	Hydrography/hydrology (surface and subsurface flows and levels), major and flash floods, glaciers; ocean levels and tides; soils geology: rock types/ages/properties/structure, landslides, underground caves; meteorology and climatology; archaeology; seismology: active faults, seismicity, seismic wave propagation, ground motion; volcanology; disasters: earthquakes, hurricanes, tornadoes, drought, storms, snow, fires, falling meteorites, etc.
Engineering data	Control structures: locks, dams, levees, huge retaining walls; pump stations; building inventories/codes; transportation (cars, metros, trains, boats, etc.), bridges, tunnels, airports, cable cars; utility infrastructure, petrol pipelines, power lines, gas pipes, electricity cables; hazardous plants: chemical plants; nuclear plants, gas stations, etc.; hazardous storage: ammunition and weapon storage; petrol storage depot, etc.; hazardous transportation; critical facilities: huge sport grounds, huge theatres, etc.; communication systems; computer crashes; etc.
Environmental data	Hazardous sites; water quality; critical facilities, etc; wars, terrorism attacks, huge demonstrations (sometimes), etc.
Economic data	Demography; employment and services locations; epidemiology, transmissible illness, etc.
Response data	Evacuation routes; management plans; aircraft routes; personnel deployment; equipment deployment; warning system; hospitals; shelters, etc.

11.4.3   Structuring the system

An example (Figure 11.10) of operational concepts for structuring a system dealing with urban risks was recently given by Cochran and Power (1999). The four concepts, Monitor, Assess, Plan and Act are organised cyclically. Facing a threat, we have to monitor it especially by collecting data and studying them. Doing so, we can analyse various local situations, and when the threat is escalating, we need to plan, namely to make scenarios in order to diminish this threat. Finally an action plan is made in order to solve the threat. Together with this plan, evaluations are made in order to draw lessons for preventing a new threat. Figure 11.11 emphasises the relationships between the actors and the systems in order to solve the problem of risk management.

11.4.4 Urban risk mapping and GIS

Starting from the previous analysis, once the data are stored in the GIS, one very important step is to map urban risks, in order to show the local councillors and the urban decision makers the more risky zones, their risk factors, and to establish the priorities. An example of an urban risk zone is given Figure 11.12, which was established for the Italian city of Ravenna under a more pessimistic scenario. This map comes from a research project of the University of Padua, Italy for the risks along the Adriatic Sea. Visit the following site for details: http://cenas.dmsa.unipd.it/results/littoral/littoral.html.

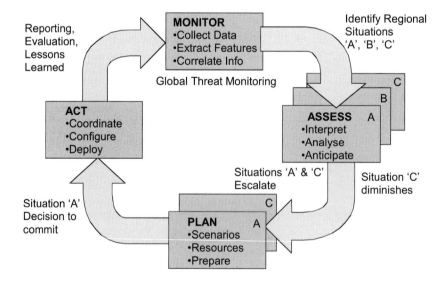

Figure 11.10 The operational concepts of an urban risk system.
After Cochran and Power 1999, published with permission.

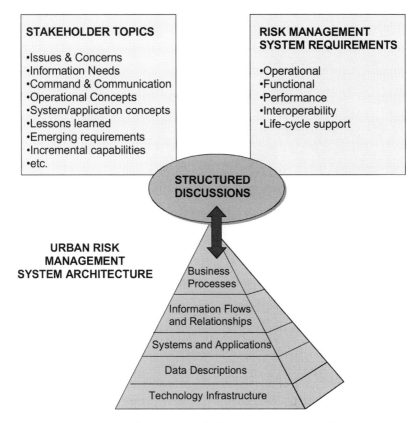

STAKEHOLDER TOPICS

•Issues & Concerns
•Information Needs
•Command & Communication
•Operational Concepts
•System/application concepts
•Lessons learned
•Emerging requirements
•Incremental capabilities
•etc.

RISK MANAGEMENT
SYSTEM REQUIREMENTS

•Operational
•Functional
•Performance
•Interoperability
•Life-cycle support

STRUCTURED
DISCUSSIONS

URBAN RISK
MANAGEMENT
SYSTEM ARCHITECTURE

Business
Processes

Information Flows
and Relationships

Systems and Applications

Data Descriptions

Technology Infrastructure

Figure 11.11 Relationships between stakeholders, requirements and system architecture via structure discussions.
After Cochran and Power, 1999, published with permission.

11.4.5   *Information to citizens concerning risks*

When a risk is approaching, one solution is to describe everything through the web; of course this kind of system must be very robust *vis-à-vis* electricity shortage and telecommunication problems. To illustrate the possibility of a real-time system for risk management and control, see the following example in the Italian city of Genoa, at this address: http://www.comune.genova.it/ protezioneciv/main_e.htm in which they mainly consider two types of urban risks, floods and landslides. Figure 11.13 inspired from the previous Genoa site, gives the structure of an Internet site targeted at citizens.

If it seems easy to inform the dwellers and the people working in the city by using the web, it is not so easy to inform people just traversing the city, or who are temporarily in the city. An interesting solution is to couple GPS to all portable computers so that they can be reached via spatial mail. Let me take

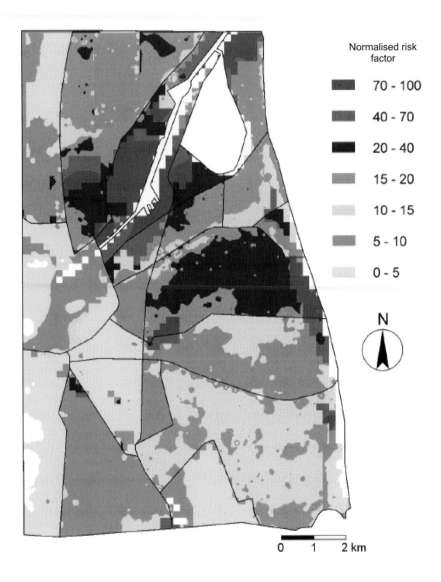

Normalised risk
factor

▮ 70 - 100

▮ 40 - 70

▮ 20 - 40

▮ 15 - 20

▮ 10 - 15

▮ 5 - 10

▮ 0 - 5

N

0   1   2 km

Figure 11.12 Risk map of the City of Ravenna, Italy for 2100, with the pessimistic
          scenario and a 10-year event.
Source: http://cenas.dmsa.unipd.it/results/littoral/fig30.html. 'CENAS: Coastline
Evolution of the Upper Adriatic Sea due to Sea Level Rise and Natural and
Anthropgenic Land Subsidence' G. Gambolati (ed.) p. 342 © 1998 by Kluwer Academic
Publishers. Reproduced with kind permission from Kluwer Academic Publishers.

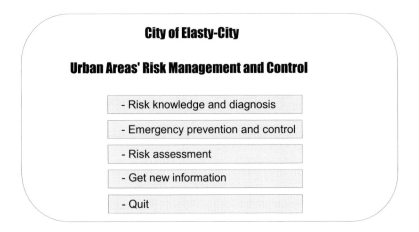

Figure 11.13 Home page of an extranet system for information to citizens for urban areas' risk management and control.

an example. Assume that we want to inform all persons within a sector of 10-mile radius from the city hall. In this case, a mail can be targeted to all computers located within this zone. For generalising this technique, a new kind of electronic address must be defined, that is to say using GPS co-ordinates in addition to computer IP number.

11.5   Example of a telegeomonitoring system for hazmat transportation planning[5]

The city and region of Mohammedia, Morocco has intensive chemical industry activities. The economic life of the region is linked to its geographical position; first due to the proximity of the Moroccan economic capital Casablanca, and second, due to the presence of the oil and chemical industries installed to the edge of Atlantic Ocean of the region at the port of Mohammedia. Industrial installations receive and dispatch many harmful materials that present, in case of accidents, risks to people and the environment. To cope with this vulnerability, a research project was initiated to develop a spatial information system to deal with the management of risks and the routeing of hazardous materials for transportation. This system makes reference to the Geographical Information Systems (GIS) technology and to the Decision Support System (DSS).

---

5   A preliminary version of this section was published in Boulmakoul A., Laurini R., Servigne S. and Idrissi M.A.J. (1999) 'First Specifications of a TeleGeoMonitoring System for the Transportation of Hazardous Materials.' In *Computers, Environment and Urban Systems*, vol. 23, 4, July 1999, pp. 259–70.

Telecommunication aspects present an important dimension for this project, and especially the contribution of the GPS and Internet technology. The proposed system will allow the analysis of accidental impact scenarios (real or simulated), and to give telemonitoring services for the transportation of hazardous materials. The telemonitoring system aims to allow public operators to integrate better the new computer technologies concerning risk management and to operate a vigilant control for trucks that haul hazardous materials.

11.5.1   Architectures

The object of this section is to describe the technical architecture of the hazardous material transportation monitoring system. This architecture shows two fundamental control types entities:

- the primary control centre (PCC) is a grouping node (only one PCC for a region to control)
- the secondary control centre (SCC) (there can be several). Each secondary node is dependent upon a geographical zone under the control of a PCC (Figure 11.14).

Obviously, this decomposition implies a controlled communication flow management. The SCC nodes can be considered as distributed agents for the supervision of hazardous materials transportation. The PCC is a main node that centralises the event's trackability, and files itineraries provided by routeing techniques.

The totality of nodes are federated by an Intranet protected network, and opened toward the Internet. This private network system guarantees authoritative control services and supervision, and the Extranet system allows professional users to access services given by nodes controllers. The control nodes include a totality of applications (see Figure 11.15) that we can summarise below.

- A geographical information system (GIS) materialises the geographical facet of the decision support system. This component integrates spatial data of the city of Mohammedia: civil infrastructure, industrial zone, urban network, port of energy, oil refinery oil, chemical factories, land use, accident data on hazardous materials transportation, etc. Some hypermedia functionalities have also been incorporated into the GIS.
- Two relational databases store the resources of each node; a multimedia database of hazardous materials, and a regulation/textual database that manage the regular aspect of hazardous materials.
- The spatial decision support system holds tools of assistance with the decision (routeing according to some criteria) and spatial analysis. This system uses a GIS and the various databases.

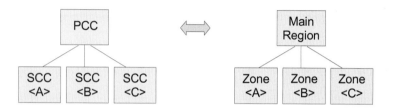

Figure 11.14 Zones and controllers hierarchy.

- The communication server feeds the spatial decision support system with the positions of trucks, and to deduce in real time, itineraries with lesser risks. The communication server feeds archives of PCC itineraries, and keeps the history of events. This server also allows authorised users (police) access to the PCC or SCC resources.
- The Web server offers JAVA applets, allowing access to the various node resources. The navigator begins these applications in Internet and in Intranet.

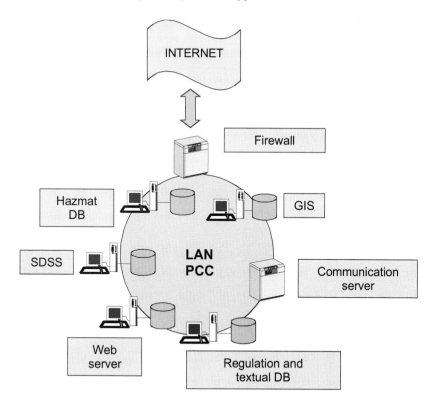

Figure 11.15 Global architecture of hazmat monitoring system: primary control centre viewpoint.

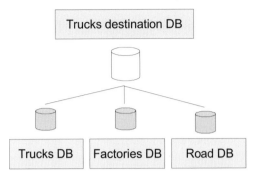

Figure 11.16 Databases hierarchy.

To communicate with the client and the server applications, a 'middleware' bridge-like system is used in a transparent manner. This suitable bridge transports SQL statements and their results (standard ODBC/JDBC[6]).

The primary node keeps a global view of the region to be supervised (global urban network of the region). However the secondary nodes store the description of geographical zones under their control. Information collected by secondary controllers will be filed at this level. Secondary controllers contribute to applications of the central web server, to reply to user requests. The primary controller unifies the global image of the network for the maintenance of the global consistency of the monitoring system. The itinerary database (Figure 11.16) at the level of the primary controller displays three databases namely for factories, trucks, and road information.

The secondary controller (Figure 11.17) is directly linked to the communication with trucks. It has responsibility for communications processing in two directions. The guidance of the trucks by dynamic routeing is made locally, whereas important events (accidents, special hazardous materials) are reported to the primary controller.

11.5.2 Toward a telegeomonitoring object components specification

From this structure, we can suggest that a real-time GIS is a particular case of a telegeomonitoring system with the following components:

- GIS engine, that is specialised for operations and spatial manipulations
- data server (data engine) whose role is to provide values to attributes of the application objects, these data coming from several data sources such as sensors, remote databases, files, remote simulators, etc.

---

6   See for instance *Database Programming With Jdbc and Java* by George Reese (1997) O'Reilly & Associates, 240 p.

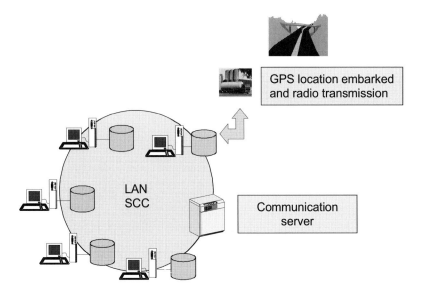

Figure 11.17 Technical network infrastructure: secondary control centre viewpoint.

• database engine, that represents the textual DBMS aspects. It operates with the GIS engine to achieve GIS functions.

11.5.3   Data collection process

The client/server mechanism is used in the telegeomonitoring context, a server process on the side of the Data Server and several processes to the level of interfaces with data sources. The main reason for using this paradigm is its great flexibility for the specific development of data collection from sensors. This practice ensures the modularity of communications programs. For the other forms of communications the CORBA/ACTIVEX[7] support offers guarantees of opening and interoperability between software components. Figure 11.18 illustrates the two levels of the data collection process.

11.5.4 Hazardous materials transportation routeing

The principal objective is to develop a 'Graph Toolbox' integrated in a spatial decision support system named SADS, which in general consists of the various navigation operators of graphs in general, and fuzzy graphs in particular (Delgado *et al.* 1990). The use of a geographical information system to model

---

7   See for instance *Designing and Using ActiveX Controls; With CD-ROM* by Tom Armstrong (1997) IDG Books Worldwide, 512 p.

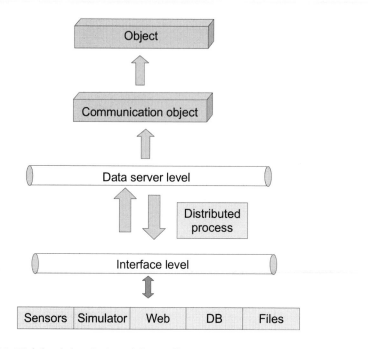

Figure 11.18 High-level description of data-collection process.

the road infrastructure, is essential for the supervision of hazardous materials transportation. SADS makes it possible to evaluate the routeing at the lesser risk for a hazardous material vehicle.

11.5.5   Risk modelling

The guide of the American Department of Transport written by the Federal Highway Administration (DOT 1989) gives the procedures for calculation of risk per segment of route. These calculations are essential for routeing algorithms that will identify minimum risk routes, where risk is established as being the product of the probability of having an accident by the consequences in term of cost, which can be expressed on the route segment in question and on its vicinity:

$$\text{Risk} = \text{Accident probability} \times \text{Accident consequences}$$

The risk calculation supposes the existence of the probabilities of occurrence of accidents on a route section. However, information is often insufficient to allow this calculation. In certain cases, the absence of data results in taking null probabilities, by considering that the concerned sections are invulnerable. This remark underlines the limits of the calculation of risk according to the

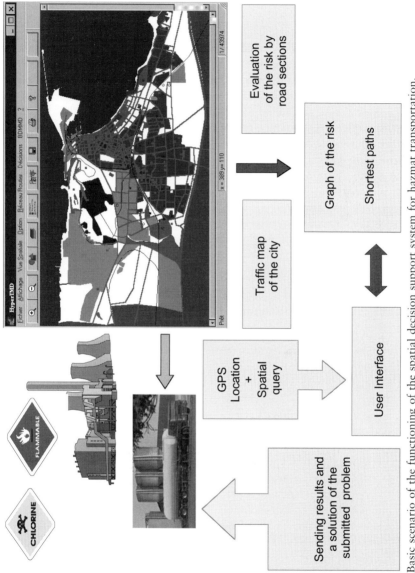

Figure 11.19 Basic scenario of the functioning of the spatial decision support system for hazmat transportation.

probabilistic method. Consequently, we have proposed a fuzzy approach to model the risk on route sections. This fuzzification of risk will allow the use of results concerning path problems by fuzzy graphs, described previously.

11.5.6 Final remarks

Software components developed in the framework of this project are generic. These templates are used currently, in the 'spatial decision support system for hazardous materials transportation planning and monitoring' project (Figure 11.19). From our experience, we can confirm the necessity to produce a new technology based on telegeomonitoring principles. Certainly, fragments constituting this technology are not original. However, their integration has to be optimised to reach the desired objectives. The hardware and software environment greatly influences global results, and especially in a real-time context. Another problem emerges underlining the absence of a methodology of a telegeomonitoring applications development. The future contribution of our practice will be to provide flexible methodological principles that we can reuse in the development of other projects.

This work implies the need to define a structure of information systems, allowing a solution to the needs of spatial data analysis (spatial data mining), and providing telegeomonitoring services for users.

11.6 Example of humanitarian assistance in urban environments

GeoWorlds (Coutinho *et al*. 1999, Kumar *et al*. 1999) is an experimental system that demonstrates how carefully integrating three key technologies can provide teams of users with a sense of shared regional vision, the ability to marshal and organise everything known about an area, displayed with respect to space and time especially for humanitarian assistance during disasters. The system seeks to provide synergy between three technologies: digital libraries, geographic information systems, and telecommunications of remote sensor data. It retrieves, organises and displays available information about a region in rich displays, allowing teams of users in distributed locations to assess situations collaboratively, develop appropriate responses, and monitor the situation's evolution. The system integrates in-house tools, telecommunication components and various products of other research institutes that are collaborating with the University of Southern California Information Sciences Institute (USC ISI).

More specifically, GeoWorlds provides the users with the ability rapidly to assemble a customised repository of information about a geographic area. It enables them to select data sets from large samples of pre-determined information stored in GIS databases, to relate the GIS data to collections of document-based information from the World

Wide Web that have been found, filtered and organised on-the-fly, and to tie these to physical events monitored by real-time sensor feeds.

The geographic and document-based information is bi-directionally correlated, i.e., by selecting a geographic region on a GIS display GeoWorlds can retrieve, filter and organise sets of documents from the Web that are associated with that region, and conversely, one can visualise in the GIS display the locations that selected documents are geo-referencing. The system has the function of a service registry which helps users find, select and initiate automated analysis tools. These tools can draw upon different sources of information (e.g., geographic and document-based information, numerical simulation of physical processes). GeoWorlds also provides an initial level of support for establishing monitoring of real-time sensor feeds in order to compare them against predictions of the analysis packages and generate warnings when actual events are not unfolding as predicted.

This functionality is provided in a framework that enables synchronous and asynchronous collaboration over finding, filtering, and organising required information and actions. The collaboration framework supports viewing and annotating the information to facilitate group decision making both to form and record conclusions about the interpretation of shared information, and to select and co-ordinate actions based upon those conclusions. The growing capability to establish monitors against the real-time data allows teams rapidly to assess situations, develop and execute responses, and – when needed – revisit their conclusions and decisions if the assumptions upon which they were based turn out to be invalid.

GeoWorlds is an investigation into multi-domain software frameworks for situation understanding and information management systems. Its current demonstration applications are in disaster response and consequence management, but the techniques are equally applicable to topics as disparate as search and rescue, weather tracking, urban planning, and natural resource management.

GeoWorlds is targeted at a community with a broad and demanding range of functional requirements; the situation understanding and information management systems community. A major challenge in designing and building such a system is not only to develop basic system capabilities but also to provide a framework where the best available tools can be integrated into the system with minimal effort and that each of these components can communicate with each other to create new applications. With this in mind, GeoWorlds was designed as a component-based system that will be able to support continuous increase of func-tionality and portability as new and more sophisticated tools become available.

The authors (Coutinho *et al.* 1999) believe that there is a broad class of applications targeted at providing total regional vision for situation understanding and management and that these applications can be grouped based on the set of functional requirements that they have in common. HA/DR (Humanitarian Assistance and Disaster Relief) operations belong to this set of applications and they consider the current version of GeoWorlds to be a reference implementation prototype supporting this kind of application.

The basic set of requirements that are identified and considered during the design of the system consist of:

- dynamic storage/retrieval, processing and analysis of data and information available about the region/situation
- simulation of physical or logistic problems involved in the situation
- visualisation and analysis of the information and results
- marking, annotating and collaboration of the information between spatially distributed teams.

Figure 11.20 sketches the organisation of components in the current GeoWorlds system. A fundamental requirement of the system is to

Functional Architecture

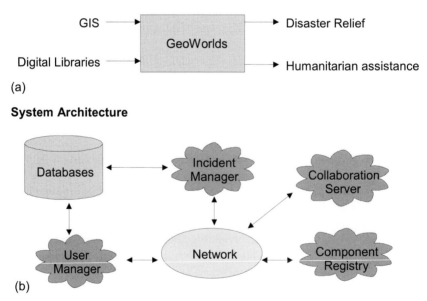

(a)

System Architecture

(b)

Figure 11.20 Architecture of GeoWorlds. (a) Functional architecture. (b) System Architecture.
From Kumar *et al.* (1999).

332

effectively provide full functional support to spatially distributed teams. The GeoWorlds system is implemented using the client-server architecture paradigm.

The current version of the system aims at helping teams performing humanitarian assistant/disaster relief operation to understand their situations in space and time. Users can rapidly create a customised repository of information about a geographic area of interest and associate it to collections of documents from the World Wide Web. Geographic and document-based information is bi-directionally correlated. The user can move from geographic to document space to learn facts about a region or to see the locations that documents are geo-referencing. These capabilities are embedded in a framework that enables synchronous and asynchronous collaboration between teams of users.

11.7   Conclusions

As urban planning is not only targeted to land-use control, but also to the welfare of citizens, the goal of this chapter was to give the reader some fresh ideas concerning urban planning for environmental control and especially urban disaster management. Now, because GIS is allied to telecommunication techniques and using Internet, new computer architectures can be offered to crisis teams and urban decision makers in order to solve old urban problems which were understudied in the past, or moreover for which it was very difficult to offer solutions. Now, it is possible by means of telegeoprocessing to solve part of them.

REFERENCES

Aalders HJGL (1999) 'The Registration of Quality in a GIS'. In Proceedings of the International Symposium on Spatial Data Quality, edited by W Shi, M Goodchild and PF Fisher, July 18–20, 1999, Hong Kong Polytechnic University, ISBN 962-367-253-5, pp 23–32.

Andrienko G. and Andrienko N (1999) 'Knowledge Engineering for Automated Map Design in Descartes'. In *Proceedings of the 7th Symposium on the Advances of GIS*, Kansas City, November 5–6, 1999, edited by C Bauzer-Medeiros, ACM-Press, pp 66–72.

ANSI/X3/SPARC (1978) *Framework Report on Database Management Systems*. American National Institute, Standards, Planning and Requirements Committee. Database System Study Group; Montvale, New Jersey, AFIPS Press.

Antenucci JC, Brown K, Croswell PL, Kevany MJ and Archer H (1991) *Geographic Information Systems: A Guide to the Technology*, Van Nostrand Reinhold.

Armstrong MP (1991) 'Knowledge Classification and Organization'. In *Map Generalization: Making Rules for Knowledge Representation*. Edited by Buttenfield BP and McMaster RB, Longman, pp 86–102.

Arnaud AM (2000) *Qualidade do Endereço na Infra-estrutura de Referenciação Espacial Indirecta*. PhD thesis, New University of Lisbon.

Arnstein SR (1969) 'A Ladder for Citizen Participation'. In *Journal of the American Institute of Planners*, vol. 35, 7, pp 216–44.

Balfanz D and Göbel S (1999) 'Bridging Geospatial Metadata Standards towards Distributed Metadata Information Systems'. Proceedings of the Third *IEEE Metadata Conference*, April 6–7, 1999, NHS, Bethesda, Maryland. Source: http://computer.org/conferen/proceed/meta/1999/papers/40/dbalfanz.html

Barquin R and Edelstein H (eds) (1997) *Planning and Designing the Datawarehouse*, Prentice Hall, 1997.

Bartelme N (1996) 'Data Structures and Data Models for GIS'. In *Data Acquisition and Analysis for Multimedia GIS*, edited by L. Mussio, G. Forlani and F. Crosilla, Springer Verlag, pp 359–76.

Batini C, Ceri S and Navathe SB (1992) *Conceptual Database Design*. Benjamin/Cummings Publishing Company, 470 pp.

Bell D and Grimson J (1992) *Distributed Database Systems*. Addison-Wesley, 409 pp.

Benslimane D, Leclercq E, Savonnet M, Terrasse MN and Yetongnon K (1999) 'On the Definition of a Generic Multi-layer Ontologies for Describing GIS Based Applications'. Accepted for publication by *Computers, Environment and Urban Systems*.

Bertin J (1967) *Sémiologie Graphique* (Paris: Gauthier-Villars) 431 pages. Translated into English by Bertin J (1983) *Semiology of Graphics*. Madison, WI, University of Wisconsin Press.

Bertin J (1983) *Semiology of Graphics*. Madison, WI, University of Wisconsin Press.

Bertino E and Martino L (1993) *Object-oriented Database Systems, Concepts and Architectures*, Addison-Wesley, 264 pp.

Bertol D (1997) *Designing Digital Space: an Architect's Guide to Virtual Reality*. John Wiley & Sons, 1997.

Bianchin A (1999) 'Steps towards a Datawarehouse for the Lagoon of Venice'. *Proceedings of the 21st Urban Data Management Symposium*, Venice, Italy, April 21–4, 1999. Source: http://www.udms.net

Blakemore M (1983) 'Generalization and Error in Spatial Databases'. In *Cartographica* vol. 21, pp 131–5.

Bobak AR (1993) *Distributed and Multi-Database Systems*. Bantam Books. 477 pp.

Bontempo C and Zagelow G (1998) 'The IBM Data Warehouse Architecture'. In *Communications of the ACM*. September 1998, vol. 41, 9, pp 39–51.

Boulmakoul A, Laurini R, Servigne S and Idrissi MAJ (1999) 'First Specifications of a TeleGeoMonitoring System for the Transportation of Hazardous Materials.' In *Computers, Environment and Urban Systems*, vol. 23, 4, July 1999, pp 259–70.

Bradshaw JM (1997) (ed) *Software Agents*, MIT Press, Cambridge 450 pp.

Brassel K, Bucher F, Stephan E-M and Vckovski A (1995) 'Completeness'. In Guptill SC and Morrison JL (eds) *Elements of Spatial Data Quality*. Oxford, Elsevier, pp 81–108.

Brodhag C (1999) *Les projets francophones dans les domaines de l'environnement et du développement durable*. St Etienne School of Mines, *Summer University for Sustainable Development and Geographic Information Systems*, July 5–9, 1999.

Brun P (1999) 'New Models of Participation New Information Tools'. *Proceedings of the 21st Urban Data Management Symposium* (UDMS), Venice, Italy April 21–3, 1999. Proceedings published in CD-ROM by the UDM Society (http://www.udms.net), paper VIII.2.

Buelher K and McKee L (1996) (eds) *The OpenGIS^{TM} Guide, Introduction to Interoperable Geoprocessing*. Open GIS Consortium. http://www.opengis.org

Cambruzzi T, Fiduccia A and Novelli L (1999) 'A Dynamic Geomonitoring System for Venetian Coastal Ecosystem: WATERS Project'. In *Computers, Environment and Urban Systems*, vol. 23, 6, November 1999, pp. 469–84.

Card SK, Mackinlay JD and Shneiderman B (1999) *Readings in Information Visualization: Using Vision to Think*, Morgan Kaufmann Publishers, 576 pp.

Caron C, Bedard Y and Gagnon P (1993) 'MODUL-R: Un formalisme individuel adapté pour les SIRS'. In *Revue de Géomatique*, vol. 3, 3, pp 83–306.

Cartwright W (1999) 'Extending the map Metaphor Using Web Delivered Multimedia'. In *International Journal of Geographical Information Science*, vol. 13, 4 June 1999 pp 335–53.

Castells M (1977) *The Urban Question*. English edition. Edward Arnold.

Chadwick G (1971) *Systems View in Planning*. Oxford: Pergamon Press

Chandrasekaran B and Josephson JR (1999) 'What are Ontologies, and Why do we Need them?' In *IEEE Intelligent Systems*, January/February 1999, pp 20–6.

Chen C (1998) 'Bridging the Gap: The Use of Pathfinder Networks in Visual Navigation'. In *Journal of Visual Languages and Computing*, vol. 9, 3, June 1998, pp 267–86.

Chen C (1999) *Information Visualisation and Virtual Environments*. Springer-Verlag, London. ISBN: 1-85233-136-4

Chen P (1976) 'The Entity-relationship Model, Toward a Unified View of Data'. In *Transactions on Database Systems*, 1, 1, March 1976, pp 9–35.

Chen X (1999) 'Range Image Analysis for 3D-City Modelling in Kyoto-City'. In *Proceedings of the International Workshop on Urban 3D/multi-Media Mapping*

(UM3'99), Tokyo, September 30–October 2, 1999, edited by R Shibasaki and Z Shi, pp 111–118.

Ciciarelli JA (1991) *A Practical Guide to Aerial Photography, with an Introduction to Surveying.* Van Nostrand Reinhold, 261 pp.

Cochran LE and Power M (1999) *Architectural Framework for Disaster Management Systems,* Document # FMT-980633–01, May 1999, Canada Centre for Remote Sensing, NRCAN-98–0633.

Codd E (1970) 'A Relational Model for Large Shared Data Banks'. In *Communications of the Association for Computing Machinery,* 13, 6 June 1970, pp 377–87.

Coleman DE and Khanna R (1995) (eds) *Groupware: Technology and Applications.* Prentice Hall, 576 pp.

Coleman DJ and LI S (1999) 'Developing a Groupware-based Prototype to Support Geomatics Production Management'. In *Computers, Environment and Urban Systems,* vol. 23, 4 July 1999, pp 315–31.

Collin C (1995) *Risques Urbains.* Editions Continent Europe, 224 pp.

Conklin J and Begeman ML (1988) 'gIBIS: a Hypertext Tool for Exploratory Policy Discussion'. In *ACM Transactions on Office Information Systems,* 6, 4, pp 303–31.

Connor DM (1992) *Constructive Citizen Participation: A Resource Book,* fourth edition; Development Press, 5096 Catalina Tce, Victoria, B.C, V8Y 2A5; 220 pp. http://www.islandnet.com/~connor/resource_index.html

Cortopassi-Lobo M and Guetter AK (1999) *Real-time GIS Application for Environmental Planning and Flooding Risk Prevention for the State of Parana in Brazil.* First International Workshop on TeleGeoProcessing, Lyon, France, May 6–7, 1999, edited by R. Laurini, pp. 118–24.

Coulet P and Givone P (1995) 'Hypermedia pour la gestion des informations sur les inondations'. In *Revue Internationale de Géomatique,* vol. 5, 2 pp 225–44.

Coulet P, Laurini R and Givone P (1995) 'Conception d'un hypermedia pour la gestion des informations sur les inondations d'un bassin versant'. In *Actes des 2ièmes Journées de la Recherche* edited by R. Jeansoulin, CASSINI, Marseille, 15–17 Novembre, 1995. Marseille: LIM, 1995. 10 pp.

Coutinho M, Neches R, Bugacov A, Yao KT, Kumar V, Ko IY and Eleish R (1999) *GeoWorlds: a Geographically-based Information System for Situation Understanding and Management.* First International Workshop on TeleGeoProcessing, Lyon, France, May 6–7, 1999, edited by R. Laurini, pp 16–22. To be published by Geo-Informatica.

Cowen DJ, Shirley WL and Jensen J (1998) 'Collaborative GIS: A Video-conferencing GIS for Decision Makers'. *Proceedings of the International Conference on Geographic Information, Lisbon,* September 7–11, 1998. CD-ROM produced by Imersiva: www.imersiva.ch.

Craig WJ (1998) 'The Internet Aids Community Participation in the Planning Process'. *Proceedings of the COST-UCE C4 International Workshop on Groupware for Urban Planning,* Lyon, February 4–6, 1998. Edited by R. Laurini, pp 10.1–11.10. Republished in *Computer, Environment and Urban Systems,* vol. 22, 4 pp 393–404.

Crowther P (1999) *The Nature and Acquisition of Expert Knowledge to be Used in Spatial Expert Systems for Classifying Remotely Sensed Images.* PhD Thesis submitted to the University of Tasmania, Australia, 1999.

Date CJ (1987) 'Twelve Rules for a Distributed Database'. In *InfoDB,* vol. 2 Nos. 2 and 3, Summer/Fall

De Michelis G, Dubois E, Jarke M, Matthes F, Mylopoulos J, Schmidt JW, Woo C and Yu E (1998) 'A three-facet view of information systems'. In *Communication of the*

Association for Computing Machinery, vol. 41, 12, pp 64–70.

Debenham J (1998) *Knowledge Engineering: Unifying Knowledge Base and Database Design*. Springer, 465 pp.

Delgado M, Verdegay JL and Vila MA (1990) 'On valuation and optimisation problems in fuzzy graphs: a general approach and some particular cases'. In *ORSA Journal on Computing*, vol. 2, no. 1, 1990, pp 75–83

Densham PJ, Armstrong MP and Kemp KK (1995) (eds) *Collaborative Spatial Decision-Making, Scientific Report for the Initiative 17 Specialist Meeting*. NCGIA University of Santa Barbara, Technical Report 95–14, September 1995. 184 pp.

Derr K (1995) *Applying OMT: A Practical Step-By-Step Guide to Using the Object Modeling Technique*, Prentice Hall, 550 pp.

Devlin B (1997) *Data Warehouse, from Architecture to Implementation*. Addison-Wesley, 432 pp.

Dieberger A and Frank A (1998) 'A City Metaphor to Support Navigation in Complex Information Spaces'. In *Journal of Visual Languages and Computing*, vol. 9 pp 597–622.

Dieberger A and Tromp JG (1993) 'The Information City project a virtual reality user interface for navigation in information spaces'. *Proceedings of the Symposium on Virtual Reality*, Vienna, December 1–3, 1993. http://www.lc/gatech.edu/~dieberger/VRV.html

Dodge M, Doyle S, Smith A and Fleetwood S (1998) 'Towards the Virtual City: VR & Internet GIS for Urban Planning'. *Proceedings of the Seminar on Virtual Reality and Geographical Information Systems*, Birkbeck College, 22 May 1998. http://www.casa.ucl.ac.uk/publications/birkbeck/vrcity.html

DOT (1989) *Guidelines for Applying Criteria to Designate Routes for Transportation Hazardous Materials*, Report No. DOT/RSPA/OHMT-89–02, Federal Hwy. Admin, Washington, D.C.

Doubriere JC (1979) *Cours d'urbanisme appliqué*, Paris: Eyrolles, 260 pp.

Egenhofer M (1991) 'Reasoning about Binary Topological Relations'. In *Proceedings of the Second Symposium on Large Spatial Databases*, Zürich, August 28–30, 1991, LNCS 525, Springer Verlag, pp 143–59.

Elmasri R and Navathe S (1989) *Fundamentals of Database Systems*, Benjamin/Cummings Publishing Company, 802 pp.

Enemark S (1998) 'Updating Digital Cadastral Map – The Danish Experience'. *Proceedings of XXI FIG Congress*, Brighton, May 30, 1998.

Erik-Andriessen JH (1996) 'The Why, How and What to Evaluate of Interaction Technology: A Review and Proposed Integration'. In *CSCW Requirements and Evaluation*, edited by P Thomas, Springer-Verlag, 1996, pp 107–24.

Evangelatos T (1999) *Standards: Linking it together*. SORSA Workshop, Ottawa, August 10–14, 1999.

Falkner E (1995) *Aerial Mapping, Methods and Applications*. Lewis Publishers. 322 pp.

Faludi A (1973) *Planning Theory*. Oxford: Pergamon Press.

Farley J and Hecht L (1999) *OGC Discussion Paper – Disaster Management Scenarios*, Open GIS Consortium, January, 1999.

FEMA (1998) *Federal Emergency Management Agency Information Technology Architecture*, Volume 1, Version 1.0. FEMA, November, 1998.

Foot D (1981) *Operational Urban Models: An Introduction*. London: Methuen.

Foresman TW, Wiggins HV and Porter DL (1996) 'Metadata Myth: Misunderstanding the Implication of Federal Metadata Standards'. *Proceedings of the First IEEE Metadata Conference*, Silver Spring, Ma, http://computer.org/conferen/meta96/wiggins/foresman_final.html

Gahegan M (1999) 'Four Barriers to the Development of Effective Exploratory Visualisation tools for the Geosciences'. In *International Journal of Geographical Information Science*, vol. 13, 4 June 1999, pp 289–309.

Gan E and Shi W (1999) 'Development of Error Metadata Management System with Application to Hong Kong 1:20, 000 Digital Data'. In *Proceedings of the International Symposium on Spatial Data Quality*, edited by W Shi, M Goodchild and PF Fisher, July 18–20, 1999, Hong Kong Polytechnic University, ISBN 962-367-253-5, pp 396–404.

Gardner SR (1998) 'Building the Data Warehouse'. In *Communications of the ACM*, vol. 41, 9 pp 52–60.

Gauna I and Sozza A (1999) 'The Geographic Information System of the Turin City Council on the Internet'. *Proceedings of the 21st Urban Data Management Symposium*, Venice, Italy, April 21–24, 1999. Published in CD-ROM. Source: http://www.udms.net

Glance NS, Pagani DS and Pareschi R (1996) 'Generalized Process Structure Grammars (GPSG) for Flexible Representations of Work'. In Proceedings *of the ACM Conference on Computer Supported Cooperative Work*, Boston Mass, November 16–20, 1996.

Gomes de Oliveira MP, Bauzer-Medeiros E and Davis CA (1999) 'Planning the Acoustic Urban Environment: A GIS-Centered Approach'. *Proceedings of the 6th ACM Symposium on the Advances of GIS*. Edited by C Medeiros.

Goodchild M, Egenhofer M. Fegeas R and Kottman C (1999) (eds) *Interoperating Geographic Information Systems*. Kluwer Academic Press, 509 pp.

Gottsegen J (1998) 'Assessing the interests and Perceptions of Stakeholders in Environmental Debates through Argumentation Analysis'. *Proceedings of the COST-UCE-C4 International Workshop on Groupware for Urban Planning*, Lyon, February 4–6, 1998, edited by R Laurini, pp 13.1–13.12. Republished in *Computer, Environment and Urban Systems*, vol. 22, 4 pp 365–79.

Gronbaek K, Kyng M and Mogensen P (1993) 'CSCW Challenges: Cooperative Design in Engineering Projects'. In *Communications of the ACM*, vol. 36, 4, June 1993, pp 67–77.

Gruber TR (1991). 'The Role of Common Ontology in Achieving Sharable, Reusable Knowledge Bases'. In *Principles of Knowledge Representation and Reasoning*. Proceedings of the Second International Conference, edited by J Allen, R Fikes, and E. Sandewall, 1991, pp 601–2.

Gruber TR (1992). 'A Translation Approach to Portable Ontology Specifications'. In *Knowledge Acquisition*, vol. 5, 1992, pp 199–220.

Guarino, N, Carrara M and Giaretta P (1994) 'Formalizing Ontological Commitment'. In *Proceedings of National Conference on Artificial Intelligence* (AAAI-94). Seattle, Morgan Kaufmann, pp 560–7.

Guptill SC and Morrison JL (1995) (eds) *Elements of Spatial Data Quality*. Oxford, Elsevier

Hadzilacos T, Kalles D, Preston N, Melbourne P. Camarinopoulos L, Eimermacher M, Kallidromitis V, Frondistou-Yannas S and Saegrov S (2000) 'UTILNETS: A Water Mains Rehabilitation Decision Support Systems'. In *Computers, Environment and Urban Systems*, vol. 24, 3, May 2000, pp 215–32.

Hayes JR (1981) *The Complete Problem Solver*. Philadelphia Franklin Institute Press.

Hedorfer M and Bianchin A (1998) 'Un modèle structurel pour métadonnées'. In *Revue Internationale de Géomatique*, vol. 8, 1–2, pp 75–97.

Heikkila E, Moore JE and Kim TJ (1990) 'Future directions for EGIS: Applications to Land Use and Transportation Planning'. In *Expert Systems: Applications to Urban Planning*, edited by TJ Kim, LL Wiggins and JR Wright. Springer-Verlag, pp 225–40.

Henderson JA (1997) *Community Planning And Development*. Lecture notes Spring Semester, 1997, Whittier College, California, http://obie.whittier.edu/~jeffh/lecture1.html

Höllerer T, Feiner S and Pavlik J (1999) 'Situated Documentaries: Embedding Multimedia Presentations in the Real World'. In Proceedings of the IEEE *International Symposium on Wearable Computers*, San Francisco, October 18–19, 1999.

Huxhold WE (1991) *An Introduction to Urban Geographic Information Systems*. Oxford University Press, 337 pp.

Isakowitz T, Stohr EA and Balasubramanian P (1995) 'RMM: A Methodology for Structured Hypermedia Design'. In *Communications of the Association for Computing Machinery*, vol. 38, 8, pp 34–44.

Jankowski P (1998) 'Public Participation GIS under Distributed Space and Time Conditions'. *Proceedings of the COST-UCE-C4 International Workshop on Groupware for Urban Planning*, Lyon, February 4–6, 1998, edited by R. Laurini, pp 11.1–11.9.

Johansen B, Sibbet D, Benson S, Martin A, Mittman R and Saffo P (1991) *Leading Business Teams: How Teams Can Use Technology and Group Process Tools to Enhance Performance*. Addison-Wesley, Reading Mass.

Johnson B and Shneiderman B (1991) 'Treemaps: a Space-filling Approach to the Visualisation of Hierarchical Information Structure'. *Proc. of the 2nd International IEEE Visualization Conference*, San Diego, October 1991, pp 284–91. Reprinted in *Sparks of Innovation in Human-Computer Interaction*, edited by B Shneiderman, Ablex Publishing Corporation, 1993, pp 309–22.

Kaas M, Witkin A and Terzopolous D (1987) *Snakes: Active Contour Models*. *International Conference on Computer Vision*. London, pp 259–68.

Kang M and Servigne S (1999) 'Animated Cartography for Urban Soundscape Information'. *Proceedings of the 7th International Symposium on GIS (ACMGIS'99)*, Kansas City, November 5–6, 1999, edited by C Bauzer-Medeiros, ACM-Press, pp 116–21.

Kappel G, Rausch-Schott S and Retschitzegger (1998) 'Coordination in Workflow Management System A Rule-Based Approachs'. In *Coordination Technology for Collaborative Applications*, edited by W Conen and G Neumann. Springer Verlag, pp 99–119.

Keen, PGW (1981) 'Information Systems and Organizational Change'. In *Communications of the ACM*, vol. 24, 1, January 1981, pp 24–33.

Kim TJ, Wiggins LL and Wright JR (1990) *Experts Systems: Applications to Urban Planning*. Springer-Verlag. 268 pp.

Kim W and Seo J (1991) 'Classifying Schematic and Data Heterogeneity in Multidatabase Systems'. In *Computer*, December 1991, vol. 24, 12, pp 12–18.

Kingston R (1998a) *Web Based GIS for Public Participation Decision Making in the UK*. National Center for Geographic Information and Analysis. *Proceedings of the Workshop on Empowerment, Marginalisation, and Public Participation GIS*, Santa Barbara, Ca, October 14–17, 1998. http://www.ccg.leeds.ac.uk/vdmisp/publications/sb_paper.html

Kingston R (1998b) 'Accessing GIS over the Web: an aid to Public Participation in Environmental Decision Making'. *Proceedings of the Workshop of the International Association for Public Participation*, SPICE '98, Tempe Arizona, October 3–7, 1998, http:www.ccg.leeds.ac.uk/vdmisp/publications/icppit.html

Klein M (1997) 'Coordination Science: Challenges and Directions'. In *Coordination Technology for Collaborative Applications*, edited by W. Conen and G. Neumann. Springer Verlag, pp 161–76.

Kraak MJ and MacEachren AM (1994) 'Visualisation of the Temporal Component of spatial Data'. In *Advances in GIS Research*. Edited by TC Waugh. *Proceedings of the 6th International Symposium on Spatial Data Handling*, London, Taylor and Francis, pp 391–409.

Kraak MJ and Ormeling FJ (1996) *Cartography, Visualisation of Spatial Data*, Addison Wesley Longman, 222 pp.

Kraak MJ and van Driel R (1997) 'Principles of Hypermaps'. In *Computers and Geosciences* 1997, vol. 23, no. 4, pp 457–64; http://www.itc.nl/~kraak/hypermap/

Kraus K (1993) *Photogrammetry, Fundamentals and Standard Processes*, Volume 1. Ummler/Bonn.

Kruijff E (1998) *Moving Sketches, Designing and Communicating Preliminary Design Ideas*, Graduation Thesis, Utrecht University, The Netherlands, June 1998 http:// kwetal.ms.mff.cuni.cz/~ernst/moving/vr_arch.htm

Krygier JB (1994) 'Sound and Geographic Visualization.' In *Visualization in Modern Cartography* edited by AM MacEachren and DRF Taylor, Oxford: Elsevier Science Ltd, pp 149–66.

Kucera HA and O'Brien CD (1997) *Harmonization of the SQL/MM Spatial and TC211 Standards, Joint Canadian Contribution to ISO/TC 211 and JTC1 SC21/WG3*, CAC SC21 WG3 N429, Madrid, January.

Kumar V, Bugacov A, Coutinho M and Neches R (1999) 'Integrating Geographic Information Systems, Spatial Digital Libraries and Information Spaces for Conducting Humanitarian Assistance and Disaster Relief Operations in Urban Environments'. In *Proceedings of the 7th Symposium on the Advances of GIS*, Kansas City, November 5–6, 1999, edited by C Bauzer-Medeiros, ACM-Press, pp 146–51.

Kunz W and Rittel HWJ (1970) *Issues as Elements of Information Systems*. Technical Report 131, University of California at Berkeley.

Lahti, P (1994) 'Integrated Management of Urban Structures with New Information Technology'. *Proceedings of COST Urban Civil Engineering*, Action C4 seminar, Paris, France March 18–19, 1994, pp 88–99.

Laurini R (1978) A Control Model for Urban Systems: an Explorative Framework. In *Transactions of the Martin Centre*, vol. III, Edited by JP Steadman and J Owers, Cambridge: Woodhead-Faulkner.

Laurini R (1979) Modelling Cities as General Systems. *Sistemi Urbani*, 2, pp 27–57.

Laurini R (1980) *Contributions systémiques et informatiques au multipilotage des Villes*. State Doctorate, Claude Bernard University of Lyon, December 1980.

Laurini R (1982a) 'French Local Planning Practice'. In *Systems Analysis in Urban Policy-Making and Planning*. Edited by M Batty and B Hutchinson. Plenum Press 1982, pp 193–203.

Laurini R (1982b) 'Nouveaux outils informatiques pour l'élaboration conjointe des plans d'urbanisme'. *Proceedings of the 9th European Symposium on Urban Data Management Symposium* (UDMS) Valencia, Spain, October 26–29, 1982.

Laurini R (1989) 'Guest editorial'. In *Sistemi Urbani*, Special Issue on Expert Systems for Urban Planning. n XI, 4, April 1989.

Laurini R (1994a) 'Multi-Source Updating and Fusion of Geographical Databases'. In *Computers, Environment and Urban Systems*, vol. 18, 4, November 1994, pp 243–56.

Laurini R (1994b) 'Sharing Geographic Information in Distributed Databases'. *Proceedings of the Annual Conference URISA (Urban and Regional Information Systems Association)* Milwaukee, August 7–11, 1994, pp 441–55.

Laurini R (1995). 'Computer Supported Cooperative Work'. In *Cognitive Aspects of Human-Computer Interaction for GIS*, edited by TL Nyerges, DM Mark, R

Laurini and MJ Engenhofer, Kluwer Academic Publishers, NATO ASI Series D, vol. 83, 1995, pp 285–6.

Laurini R (1996) 'Raccordement géométrique de bases de données géographiques fédérées'. In *Ingénierie des Systèmes d'Information*, vol. 4, 3, pp 361–88.

Laurini R (1998a) 'Groupware for Urban Planning: An Introduction'. In *Computers, Environment and Urban Systems*, vol. 22, 4 July 1988, pp 317–33.

Laurini R (1998b) 'Spatial Multi-database Topological Continuity and Indexing: a Step Towards Seamless GIS Data Interoperability'. In *International Journal of Geographical Information Science*, vol. 12, 4, June 1998, pp 373–402.

Laurini R (2000) 'An Introduction to TeleGeoMonitoring: Problems and Potentialities'. In *GIS Innovations*, edited by P Atkinson and D Martin, pp 11–26, Taylor and Francis, 1999.

Laurini R and Milleret-Raffort F (1989) *L'ingénierie des connaissances spatiales*. Paris, Hermès.

Laurini R and Milleret-Raffort F (1990) 'Principles of Geomatic Hypermaps'. *Proceedings of the 4th International Symposium on Spatial Data Handling*. Zürich, 23–27 Juillet 90, Edited by K Brassel, pp 642–651.

Laurini R and Milleret-Raffort F (1991) 'Using Integrity Constraints for Checking Consistency of Spatial Databases'. *Proceedings of the GIS/LIS'91* Conference, Atlanta, Georgia, October 28–November 1, 1991, pp 634–42.

Laurini R and Milleret-Raffort F (1994) 'Distributed Geographical Databases: some Specific Problems and Solutions'. *Proceedings of the 7th International Conference on Parallel and Distributed Computing Systems*. Las Vegas, October 5–8, 1994, pp 276–83.

Laurini R and Milleret-Raffort F (1995) 'Indexation spatiale dans une fédération de bases de données géographiques'. In *Revue Internationale de Géomatique*, vol. 2 1995, pp 245–68.

Laurini R and Servigne S (1997) 'An Information System for Urban Soundscape', *Proceedings of URISA '99*, Chicago, August 21–25, 1999, pp 736–46.

Laurini R and Thompson D (1992) *Fundamentals of Spatial Information Systems*. Academic Press, 680 pp.

Laurini R and Vico F (1999) '3D Symbolic Visual Simulation of Building Rule Effects in Urban Master Plans', *Proceedings of the Urban 3D/Multi-Media Mapping* (UM3'99) conference, Tokyo, September 30–October 2, 1999, edited by R Shibasaki and Z Shi, pp 33–40.

Lbath A (1997) *AIGLE, un environnement visuel pour la conception et la génération automatique d'applications géomatiques*. PhD thesis INSA Lyon, November 1997.

Lbath A, Aufaure-Portier MA and Laurini R, (1997) 'Using Visual Language for the Design and Query in GIS Customization'. *Procedings of the 2nd International Conference on Visual Information Systems*, San Diego, December 1997, pp 197–204.

Leavitt HJ (1965) 'Applying Organizational Change in Industry: Structural, Technological and Humanistic Approaches'. In *Handbook of Organizations*, edited by J March. Chicago, IL: Rand McNally.

Leick A (1995) *GPS Satellite Surveying*. John Wiley and Sons, 560 pp.

Lemoigne JL (1977) *La théorie du système général. Théorie de la modélisation*. Presses Universitaires de France.

Lester M and Chrisman N (1991) 'Not All Slivers Are Skinny: A Comparison of two Methods for Detecting Positional Errors in Categorical Maps'. *Proceedings of GIS/LIS'91* Atlanta conference, October 28–November 1, 1991, vol. 1, pp 648–57.

Livingston G and Rumsby B (1997) 'Database Design for Data Warehouses: The Basic Requirements'. In *Planning and Designing the Datawarehouse*, edited by R Barquin and H Edelstein, Prentice Hall, 1997, pp 179–98.

Longley P, Goodchild MF, Maguire D and Rhind DW (1999) *Geographic Information Systems, Principles and Techniques*. Wiley, 1100 pp.

Lynch PJ and Horton S (1997) 'Imprudent Linking Weaves a Tangled Web'. In *Computer*, July 1997, pp 115–17.

Maack U (1999) 'Shared Updated and Maintenance of Spatial Objects in a Municipal GIS'. *Proceedings of the SORSA Workshop*, Ottawa, August 11–14, 1999.

MacEachren AM (1994) 'Visualization in Modern Cartography: Setting the Agenda'. In *Visualization in Modern Geography*, edited by AM McEachren and DRF Taylor, Oxford, Elsevier Science Ltd, pp 1–12.

Malhing DE, Craven N and Croft WB (1995) 'From Office Automation to Intelligent Worflow Systems'. In *IEEE Expert Systems*, June 1995, pp 41–7.

Marx RW and Saalfeld AJ (1988) *Programs for Assuring Map Quality at the Bureau of the Census*. US Dept of Commerce, Bureau of the Census, March 1988.

Maurer F and Pews G (1996) 'Supporting Cooperative Work in Urban Land-use Planning'. *Second International Conference on the Design of Cooperative Systems* (COOP'96), June 12–14, 1996, Juan-les-Pins.

May AD, Mitchell G and Kapiszewska D (1996) 'The Quantifiable City: the Development of a Modelling Framework for Urban Sustainability Research'. In *Information Systems and Processes for Urban Civil Engineering Applications*, edited by Ugo Schiavoni, Rome, November 21–22, 1996. Published by the European Communities in 1998, European Commission EUR 18325 EN, ISBN 92-828-3734-3. pp 124–42.

McKeown DM, Harvey WA and Wixson LE (1989) 'Automating Knowledge Acquisition for Aerial Image Interpretation'. In *Computer Vision and Graphics*, vol. 46, pp 37–81.

Müller JC and Laurini R (1997) 'La cartographie de l'an 2000'. In *Revue Internationale de Géomatique*, vol. 7, 1, pp 87–106.

Murao M, Arikawa M and Okamura K (1999) 'Augmented/Reduced Spatial Hypermedia Systems for Networked Live Videos on Internet'. In *Proceedings of the International Workshop on Urban 3D/multi-Media Mapping* (UM3'99), Tokyo, September 30–October 2, 1999, edited by R Shibasaki and Z Shi, pp 15–20.

Neal DC (1997) 'How to Justify the Datawarehouse and Gain Top Management Support'. In *Planning and Designing the Datawarehouse*, edited by R Barquin and H Edelstein, Prentice Hall, 1997, pp 91–115.

Negoita CV and Ralescu D (1975) *Applications of Fuzzy Sets to Systems Analysis*, Birkhäuser Verlag.

Nijkamp P and Scholten JH (1991) 'Spatial Information System: Design, Modelling and use in Planning', in *Proceedings of the 5th Forum of URSA-NET Computers in Spatial Planning*, Patras, June 7–9, 1991, edited by N. Polydorides, pp 13–24.

Nobre JA (1999) 'Improving Community Participation in Urban Planning Decision' Making. Private communication, Public Works Ministry of the Macau Government, May 1999.

Nunamaker JF, Briggs RO and Mittleman DD (1995) 'Electronic Meeting Systems: Ten Years of Lessons Learned'. In *Groupware: Technology and Applications*, edited by Coleman D and Khanna R, Prentice Hall, 1995, pp 149–92.

Nyerges TL, Barndt M and Brooks K (1997) 'Public Participation Geographic Information Systems'. *Proceedings AutoCarto 13* Vol. 5 APSRS, Seattle, Washington, pp 224–33.

O'Brien D (1999) 'Data Access: A Common Window', *Proceedings of the SORSA Workshop*, Ottawa, August 11–14 1999.

Odell JJ and Fowler M (1998) *Advanced Object-Oriented Analysis and Design Using UML*. SIGS Books/Cambridge Press, June 1998, 264 pp.

O'Leary DE (1998) 'Using AI in Knowledge Management: Knowledge Bases and Ontologies'. In *IEEE Intelligent Systems and Their Applications*, vol. 13, 3, May/June 1998 pp 34–9.

Oestereich B and Cestereich B (1999) *Developing Software with UML*, (The Addison-Wesley Object Technology Series), Addison-Wesley, 352 pp.

Oszu T and Valduriez P (1989) *Principles of Distributed Database Systems*. Englewood Cliffs (New Jersey), Prentice Hall. Second edition in 1999.

Palmer JD, Fieds NA and Brouse PL (1994) 'Multigroup Decision-Support Systems in CSCW'. In *Computer*, vol. 27, 5, May 1994, pp 67–72.

Pantazis DN (1994) *Methodological Analysis for Design and Development of GIS*, Phd Thesis, Université de Liège, Belgium, December 1994, 556 pp.

Peinel G and Rose T (1999) 'Graphical Information Portals, the Application of Smart Mapsfor Facility Management in GEONET 4D'. In *First International Workshop on TeleGeoProcessing*, Lyon, France, May 6–7, 1999, edited by R Laurini, pp 6–15.

Peterson MP (1995) *Interactive and Animated Cartography*. New Jersey, Prentice Hall, pp 464.

Plümer L (1996) 'Achieving Integrity of Geometry and Topology in GIS'. In *Proceedings of the First International Conference on GIS in Urban and Environmental Planning*, edited by T Sellis and D Georgoulis, April 1996, pp 45–60.

Qiu J and Hunter G (1999) 'Managing Data Quality Information'. In *Proceedings of the International Symposium on Spatial Data Quality*, edited by W Shi, M Goodchild and PF Fisher, July 18–20, 1999, Hong Kong Polytechnic University, ISBN 962-367-253-5, pp 384–95.

Raden N (1997) 'Choosing the Right OLAP Technology'. In *Planning and Designing the Datawarehouse*, edited by R Barquin and H Edelstein, Prentice Hall, 1997, pp 199–224.

Ram S (1991) 'Heterogeneous Distributed Database Systems'. In *Computer*, December, vol. 24, 12, pp 7–10.

Rinner C (1999) 'Argumaps for Spatial Planning'. In *Proceedings of the First International Workshop on TeleGeoProcessing*, edited by R Laurini, Lyon, May 6–7, 1999, pp 95–102.

Roberts D (1999) 'Issues in GIS Data Maintenance'. In *Proceedings of the URISA Conference*, edited by MJ Salling, August 21–25, 1999, Chicago, pp 237–46.

Rouhani S and Kangari R (1987) 'Landfill Site Selection: A Microcomputer Expert System'. In *International Journal for Microcomputers in Civil Engineering*, vol. 2, 1, March 1987, pp 47–53.

Rumbaugh J, Blaha M, Lorensen W, Eddy F and Premerlani, W (1991) *Object-Oriented Modeling and Design*, Prentice-Hall, Inc. 528 pp.

Rydin Y (1998) *Urban and Environmental Planning in the UK*. MacMillan Press Ltd. 399 pp.

Sarjakoski T (1998) 'Networked GIS for Public Participation Emphasis on Utilizing Image Data'. *Proceedings of the COST-UCE C4 International Workshop on Groupware for Urban Planning*, Lyon, February 4–6, 1998. Edited by R Laurini, pp 9.1–9.13. Republished in *Computers, Environment and Urban Systems*, vol. 22, 4 pp 381–92.

Sarly RM (1972) *The Planning Process*. University College London Working Paper.

Schäl T (1996) *Workflow Management Systems for Process*. Springer Verlag. Lecture notes in computer science 1096. 205 pp.

Schmidt-Beltz B, Rinner C and Gordon TF (1998) 'GeoMed for Urban Planning First User Experiences'. *Proceedings of the 6th International ACM Symposium on*

Advances in Geographic Information Systems (ACM-GIS'98), Washington DC, November 6–7, 1998. Edited by R Laurini, K Makki and N Pissinou, pp 82–5.

Schuler P (1996) *New Community Networks: Wired for Change*. Addison-Wesley Longman, 1996, 528 pp.

Sena JA and Shani AB (1999) 'Intellectual Capital and Knowledge Creation: Towards an Alternative Framework'. In *Knowledge Management Handbook*, edited by J Liebowitz, CRC Press, pp 8.1–8.16.

Servigne S (1993) *Base de données géographiques et photos aériennes: de l'appariement à la mise à jour*. PhD Dissertation, Institut National des Sciences Appliquées de Lyon, February 1993.

Servigne S, Laurini R and Rumor M (1991) 'PHOTOPOLIS Project: Confronting Pictorial and Geographical Objects'. *Proceedings of the 14th Urban Data Management Symposium*, Odense, Denmark, 29–31 May 1991, pp 245–55.

Sheth AP and Larson JA (1990). 'Federated Database Systems for Managing Distributed, Heterogeneous, and Autonomous Databases'. In *Computing Surveys*, 22, 3, pp 183–236.

Shiffer MJ (1992) 'Towards a Collaborative Planning System'. In *Environment and Planning B: Planning and Design*, vol. 19, pp 709–22.

Shiffer MJ (1995) 'Issues of Collaborative Spatial Decision-Support in City Planning'. In *Collaborative Spatial Decision-Making*, Scientific Report for the Initiative 17 Specialist Meeting edited by Densham PJ, Armstrong MP and Kemp KK. NCGIA Technical Report 95–14, September 1995. 184 pp.

Shiffer MJ (1999) 'High Tech and Public Discourse in Planning Environments'. *Proceedings of the URISA Conference*, Chicago, August 1999. See also http://yerkes.mit.edu/shiffer/MMGIS/Title.html.

Shneidermann B (1992) *Designing the User Interface*. Third Edition in 1997, Addison-Wesley Publishing Company, 600 pp.

Shneidermann B (1993) (ed) *Sparks of Innovation in Human-Computer Interaction*, Ablex Publishing Corporation, 400 pp.

Simon E (1996) *Distributed Information System, From Client-server to Distributed Multimedia*. McGraw-Hill, 1994.

Smit LA (1995) 'Rotterdam in 90 000 photos on your PC'. *Proceedings of the JEC International Conference on Geographic Information*, March 1995.

Spéry L (1999) *Historique et mise à jour de données géographiques: application au cadastre français*. PhD Dissertation, Université d'Avignon, November 1999.

Tanin E, Beigel R and Shneiderman B (1996) 'Incremental Data Structures and Algorithms for Dynamic Query Interfaces'. In *Workshop on New Paradigms in Information Visualization and Manipulation*, fifth ACM International Conference on Information and Knowledge Management (CIKM '96) (Rockville, MD, Nov. 16, 1996), pp 12–15. Also in *SIGMOD Record*, Vol. 25, 4. December 1996, pp 21–4.

Tanzi T, Guiol R, Laurini R and Servigne S (1997) 'Risk Rate Evaluation for Motorway Management'. *Proceedings of the International Emergency Management Society Conference*, (TIEMS 1997), Copenhagen, June 10–13, 1997, edited by V Andersen and V Hansen, pp 125–34. Republished in *Safety Sciences* 29, 1998, pp 1–15.

Tanzi T, Laurini R and Servigne S (1998) *Vers un système spatial temps réel d'aide à la décision*, Journées SIGURA Octobre 1997. Republished in *Revue Internationale de Géomatique* vol. 8, 3 pp 33–46.

Tanzi TJ, Laurini R and Servigne S (1999) 'A Prototype of TeleGeoMonitoring System for Road Safety and Maintenance'. *First International Workshop on TeleGeoProcessing*, Lyon, France, May 6–7, 1999, edited by R Laurini, pp 103–11.

Tardieu H and Guthmann B (1991) *Le triangle stratégique: stratégie, structure et technologie de l'information*. Paris, Editions d'Organisation, 304 pp.

Taylor DRF (1994) 'Perspectives on Visualization and Modern Cartography'. In *Visualization in Modern Cartography* Edited by AM MacEachren and DRF Taylor, Oxford: Elsevier Science Ltd, pp 333–41.

Tellez B and Servigne S (1998) 'A Methodology for Structuring Multi-Source Information: Application to Urban Database Updating with Aerial Photos'. *Proceedings of the COST-UCE C4 program*, Kiruna, Sweden, September 1998.

Thompson D, Lindsay FE, Davis PE and Wong DWS (1997) 'Towards a Framework for Learning with GIS: The Case of UrbanWorld, a Hypermap Learning Environment Based on GIS'. In *Transactions in GIS*, vol. 2, 2, pp 151–67. http://www.inform.umd.edu/geog/gis/urbanworld/html/about.html

Torres-Fonseca F and Egenhofer M (2000) 'Ontology-driven Geographic Information Systems'. In *Computers, Environment and Urban Systems*, vol. 24, 3, May 2000, pp 251–71.

Toulmin S (1958) *The Uses of Arguments*. Cambridge University Press.

Turban E and Aronson JE (1998) *Decision Support Systems and Intelligent Systems*. Prentice-Hall, fifth edition. 890 pp.

Tweed C (1998) 'Supporting Argumentation Practices in Urban Planning and Design'. *Proceedings of the COST-UCE C4 International Workshop on Groupware for Urban Planning*, Lyon, February 4–6, 1998. Edited by R Laurini. pp 6.1–6.11. Republished in *Computer, Environment and Urban Systems*, vol. 22, 4, pp 351–63.

Ubeda T (1997) *Contrôle de la Qualité spatiale des bases de données géographiques: cohérence topologique et corrections d'erreurs*. PhD, INSA de Lyon, 1 December, 1997.

Ubeda T and Egenhofer M (1997) 'Topological Error Correcting in GIS'. In *Proceedings of the 5th International Symposium on Spatial Databases*, Berlin, July 1997, edited by M Scholl and A Voisard, Lecture Notes in Computer Science, 1262, Springer Verlag, pp 283–97.

Ubeda T, Puricelli A, Servigne S and Laurini R (1997) 'A Methodology for Spatial Inconsistency Checking and Correcting'. First CASSINI *International Workshop Data Quality in Geographic Information: from Error to Uncertainty* April 21–23 1997, Paris. Hermès, pp 111–19.

Uitermark HT, van Oosterom JM, Mars NJI and Molenaar M (1999) 'Ontology-based Geographic Data Set Integration'. In *Proceedings of the Spatio-temporal Database Management Conference*, STDBM'99, Edinburgh, September 1999, edited by MH Böhlen, CS Jensen and MO Scholl, Springer Verlag, LNCS 1678, pp 60–78.

Vckovski A (1998) *Interoperable and Distributed Processing in GIS*. Taylor and Francis.

Vckovski A, Brassel KE and Schek JH (eds) (1999) 'Interoperating Geographic Information Systems', *Proceedings of the Second International Conference, INTEROP'99*, Springer Verlag, Lecture Notes in Computer Science, vol. 1580.

Verbree E, VanMaren G, Germs R, Jansen F and Kraak MJ (1999) 'Interaction in Virtual World Views – Linking 3D GIS with VR.' In *International Journal of Geographical Information Sciences*, vol. 13, 4 June 1999, pp 385–96.

Verbree E and Verzijl L (1998) 'Integrated 3D-GIS and VR, Use of Virtual Reality and 3D-GIS within the Planning Process Concerning the Infrastructure'. *Proceedings of the GISPLANET'98 conference*, Lisbon, September 1998, CD-ROM published by Imersiva http://www.immersiva.ch

Veregin H (1998) *Data Quality Measurement and Assessment*, NCGIA Core Curriculum in GIScience, posted March 23, 1998, http://www.ncgia.ucsb.edu/giscc/units/u100/u100.html.

Veregin H and Hartigai P (1995) 'An Evaluation Matrix for Geographical Data Quality'. In *Elements of Spatial Data Quality* edited by Guptill SC and Morrison JL, Elsevier, pp 167–88.

Vico F and Ottanà M (1998) 'Groupware for Urban Planning: an Italian Perspective'. *Proceedings of the COST-UCE C4 International Workshop on Groupware for Urban Planning*, Lyon, February 4–6, 1998. Edited by R Laurini, pp 14.1–14.14.

Vico F and Rossi-Doria L (1997) 'Combining 2D analyses with 3D Simulation in the Evaluation of Urban Planning Rules Effects'. In *Geographical Information 1997, from Research to Applications through Cooperation* edited by Hodgon S, Rumor M and Harts JJ (eds). Amsterdam IOS Press, vol. 2, pp 1099–108.

Vindasius D (1974) *Public Participation Techniques and Methodologies: A Resumé*. Ottawa: Information Canada.

Von Bertallanfy L (1968) *General System Theory*. New-York: George Braziller.

Watson HJ, Rainer RK and Houdeshel G (1996) *Building Executive Information Systems and Other Decision Support Applications*. John Wiley & Sons, 479 pp.

WCED (World Commission on Environment and Development) 1987. *Our Common Future: Report of the World Commission on Environment and Development*. Oxford: Oxford University Press.

Westphal C and Blaxton T (1998) *Data Mining Solutions: Methods and Tools for Solving Real-World Problems*. John Wiley & Sons, 448 pp.

Wiederhold G (1998) *Value-added Middleware Mediators*. Stanford University report, March 1998.

Wiederhold G (1999) 'Mediation to deal with Heterogeneous Data Sources'. In *Proceedings of Interoperating GIS, Interop'99*, Zürich, Switzerland, edited by A. Vckoski, K. Brassel and HJ Schek, Springer Verlag, Lecture Notes in Computer Sciences vol. 1580, pp 1–16.

INDEX